Springer Theses

Recognizing Outstanding

For further volumes:
http://www.springer.com/series/8790

Aims and Scope

The series "Springer Theses" brings together a selection of the very best Ph.D. theses from around the world and across the physical sciences. Nominated and endorsed by two recognized specialists, each published volume has been selected for its scientific excellence and the high impact of its contents for the pertinent field of research. For greater accessibility to non-specialists, the published versions include an extended introduction, as well as a foreword by the student's supervisor explaining the special relevance of the work for the field. As a whole, the series will provide a valuable resource both for newcomers to the research fields described, and for other scientists seeking detailed background information on special questions. Finally, it provides an accredited documentation of the valuable contributions made by today's younger generation of scientists.

Theses are accepted into the series by invited nomination only and must fulfill all of the following criteria

- They must be written in good English.
- The topic should fall within the confines of Chemistry, Physics, Earth Sciences and related interdisciplinary fields such as Materials, Nanoscience, Chemical Engineering, Complex Systems and Biophysics.
- The work reported in the thesis must represent a significant scientific advance.
- If the thesis includes previously published material, permission to reproduce this must be gained from the respective copyright holder.
- They must have been examined and passed during the 12 months prior to nomination.
- Each thesis should include a foreword by the supervisor outlining the significance of its content.
- The theses should have a clearly defined structure including an introduction accessible to scientists not expert in that particular field.

Armando Marino

A New Target Detector Based on Geometrical Perturbation Filters for Polarimetric Synthetic Aperture Radar (POL-SAR)

Doctoral Thesis accepted by
The University of Edinburgh, UK

 Springer

Author
Dr. Armando Marino
ETH Zurich
Institute of Environmental Engineering
HIF D24.1, Schafmattstr. 6
8093 Zurich
Switzerland

Supervisor
Prof. Iain H. Woodhouse,
The University of Edinburgh
EH8 9XP Edinburgh,
United Kingdom

ISSN 2190-5053
ISBN 978-3-642-43156-2
DOI 10.1007/978-3-642-27163-2
Springer Heidelberg New York Dordrecht London

e-ISSN 2190-5061
ISBN 978-3-642-27163-2 (eBook)

Printed on acid-free paper

Springer is part of Springer Science+Business Media (www.springer.com)

Parts of this thesis have been published in the following journal articles:

Horn, R., Marino, A., Nannini, M., Walker, N., & Woodhouse, I. H. (2008), The SARTOM Project: Tomography for enhanced target detection for foliage penetrating airborne SAR, EMRS-DTC 2008, 5th Annual Technical Conference 24–25th June, Edinburgh, UK.

Marino, A. Cloude S. R., & Woodhouse, I. H. (2009), Polarimetric target detector by the use of the polarisation fork, Proceedings on POLinSAR'09, Frascati, Roma, 2009.

Horn, R., Marino, A., Nannini, M., Walker, N. & Woodhouse, I. H. (2009), The SARTOM Project, tomography and polarimetry for enhanced target detection for foliage penetrating airborne SAR, EMRS-DTC 2009, 6th Annual Technical Conference 7–8th, Edinburgh, UK, July 2009.

Marino, A. Cloude S. R. & Woodhouse, I. H. (2009), Selectable target detector using the polarisation fork, Proceedings on IGARSS'09, Cape Town, South Africa, July 2009.

Marino, A. Cloude, S. R. & Woodhouse. I. H. (2010), A Polarimetric Target Detector Using the Huynen Fork, IEEE Transaction on Geoscience and Remote Sensing, 48, 2357–2366.

Marino, A. & Cloude S. R. (2010), Detecting Depolarizing Targets using a New Geometrical Perturbation Filter, Proceedings on EUSAR'10, Aachen, Germany, June 2010.

Marino, A. Cloude S. R. & Woodhouse, I. H. (2010), New classification technique based on depolarised target detection, Proceedings on EUSAR'10, Aachen, Germany, June 2010.

Walker, N., Horn, R., Marino, A., Nannini, M. & Woodhouse, I. H. (2010), The SARTOM Project: Tomography and polarimetry for enhanced target detection for foliage penetrating airborne P-band and L-band SAR, EMRS-DTC 2010, 6th Annual Technical Conference 13–14th, Edinburgh, UK, July 2010.

Marino, A., Cloude, S. R. & Woodhouse, I. H. (2010), Detecting Depolarizing Targets with Satellite Data: a New Geometrical Perturbation Filter, Proceedings on IGARSS'10, Honolulu, Hawaii, July 2010.

Marino, A. Cloude, S. R. & Woodhouse, I. H. (2012), Detecting Depolarizing Targets using a New Geometrical Perturbation Filter, IEEE Transaction on Geosciences and Remote Sensing, accepted.

I would like to dedicate this thesis to my grandfather, Armando Marino, with whom I share not only his name, but also very nice moments of our lifes

Supervisor's Foreword

To both detect and identify objects remotely has always been a key driver in remote sensing research, ever since the first airborne photographs. In the twentieth century, radar technology made available a whole new domain of target identification, namely the active sensing of objects with polarimetric radio waves. And when such systems made their way into Earth orbit in the 1970s, the world was then open to new possibilities of mapping and surveillance. This thesis is concerned with microwave observations using polarimetric SAR (Synthetic Aperture Radar) imaging techniques. The oft cited advantages of microwave imaging are its capability to supply useful images in almost any weather conditions and at night time, but if long enough wavelengths are used it also has the capacity to penetrate foliage.

The focus of the methods described in this thesis is on polarimetry, since a target illuminated by polarized waves generally scatters the incoming electromagnetic radiation with a different (and sometimes unique) polarization. This property makes polarimetry an invaluable tool to perform target detection/ classification. In the last decade, the community has become increasingly aware of the importance of SAR polarimetry and this has led to even more satellite systems able to acquire this type of data, including ALOS-PALSAR, RADARSAT-2 and TerraSAR-X. The availability of polarimetric modes (at least dual polarisation) in now standard on most of the future radar satellite missions.

This thesis presents a groundbreaking methodology for radar target detection. The detection approach introduced, "Perturbation Analysis", was completely novel and demonstrated a step change in the approach to interpreting polarimetric data. Perturbation Analysis was able to push the performance limits of current algorithms, allowing for the detection of targets smaller than the resolution cell and embedded in high levels of clutter. The methodology itself is extraordinarily flexible resulting in the employment of another two applications: (1) ship detection for maritime surveillance, and (2) change detection for land change analysis and classification of dynamic ecosystems. The manuscript itself is a thoroughly well-organised piece of work, covering every detail and perspective in order to provide a comprehensive vision of the various problems and solutions where PA can be employed. At the time Dr. Marino's viva, the methodology has already led to the publication of two journal papers and more than 20 conference papers. Moreover, this thesis was awarded "Best PhD Thesis 2011" by the RSPSoc (Remote Sensing and Photogrammetry Society) during their annual conference in Bournemouth, 14th of September 2011.

Prof. Iain H. Woodhouse

Preface

When I found out my thesis was going to be published I began to wonder what would make the best incipit. I started asking around for suggestions from friends with a higher experience in the publishing world. I received plenty of advice (thanks to all), however the one that caught most of my attention was: "Armando, you should start your preface by telling, in a humorous way, an experience you had during your PhD and remains impressed in your memory… in this way, surely you will capture the attention of your readers throughout all the rest [of the preface]". I have to admit that I actually evaluated the possibility that the suggestion was coming from a witty friend that just wanted to have a laugh at my preface expenses. But then, I realised that for most of life's mistakes you have to make them in order to learn from them. So, I started to ponder for something witty that would be worth the preface of this thesis. As an actual fact, I could not think of anything! Well, perhaps it is better I rephrase this last sentence. I could not think of one unique funny event because there were many episodes that got stuck in my mind, like a bottle neck.

I could tell you about the time I felt like Indiana Jones in the neither lost nor doomed German forests (hopefully, I am not breaking any copyright right now) looking for possible real targets that were giving positive detections to the algorithm. Or, I could try to describe the expressions of my roommate colleagues when they found me in one of my trances induced by the progressive rock music I would listen to whilst working. Or perhaps, I could tell about my first conference, when my friends described me as a "mental road-police waving around" to illustrate the directions of polarisations, but in actual fact I was just fighting against the feeling of running away screaming. The truth is that I find all the processes of investigation to be an incredibly fascinating game. Sometimes you win and feel like you deserve the Nobel prize for a silly calculation that a fresher could have done, and sometimes you lose and you cannot stop thinking about the return match and a new strategy that will bring you to victory. With the impossibility of finding a single episode, I cannot do more than just say that all the words, maths and simulations the reader will find in the following are the result of a game that I enjoyed a lot to play in both the good and bad moments. My ultimate hope is that the reader will enjoy the content as well. But now I am already too impatient to start telling more about the content of this thesis!

Synthetic Aperture Radar (SAR) is an active microwave remote sensing system able to acquire high resolution images of the scattering behaviour of an observed scene. In this thesis, the contribution of SAR polarimetry (POLSAR) in detection and classification of objects is described and found to add valuable information compared to previous approaches. The first two chapters will be dedicated to introducing the concepts of SAR and polarimetry that forms the basis of the following developments.

The core of this thesis is a new target detection/classification methodology that makes novel use of the polarimetric information of the backscattered field from a target and will be presented in two chapters. The first of them, Chap. 4 contains all the mathematical demonstrations aimed to bring the reader from a physical/algebraic concept to a final formula ready to apply. On the other hand, Chap. 5 is concerned with the statistical description of the detector, in order to acquire more information regarding its theoretical performances (in particular its ROC, Receiver Operating Characteristic, curve).

One unavoidable step in proposing a new algorithm to the World is its validation and in this thesis this part is taken strongly into account in two chapters. Chapter 6 proposes a validation based on data collected with an airborne system: E-SAR L-band (DLR, German Aerospace Centre) in a campaign narrowly aimed to target detection (included camouflaged conditions). Chapter 7 is concerned with satellite data, since they represents a particularly interesting scenario for target detection. The datasets includes ALOS-PALSAR L-band (JAXA, Japanese Aerospace Exploration Agency), RADARSAT-2 C-band (Canadian Space Agency) and TerraSAR-X X-band (DLR).

I hope you will enjoy this thesis as much as I did!

Acknowledgments

Surely a PhD is not the kind of task that one can accomplish without the help of a myriad of people. I am afraid that a detailed description would end up as a chronicle of the past 4 years of my life being as long as the entire thesis. It could start from my supervisor and end at the Cameo ticket boy. Therefore, considering I do not have enough time to write it nor the reader is interested in my diary, I believe there are two ways out to this problem: (1) to make the list as short as possible missing out many names; (2) to not write at all! Considering ingratitude is not amongst my favourite sins, I will attempt the first solution and try to keep the list as short as possible.

First of all I would like to thank my supervisor Dr. Iain Woodhouse for the never-ending encouragement and support he provided when pursuing my ideas, it did not matter how bizarre they appeared at first glance. Then there are my friends within the University of Edinburgh who assisted me in fitting into my new office environment and orienteering around many pubs: Iain Cameron, Karen Viergever, Bronwen Whiteney, Mahmet Karatay, Rachel Gaulton. Without them I would still be fighting with IT or secretarial nuisances and would probably ignore the real value of a good pint.

Many people contributed to the actual advance of my PhD and thesis. Amongst all I would like to give my special thanks to Shane Cloude from AEL consultants. Without his continuous suggestions and advice the entire work would not be as it appears in this thesis. I would like to thank all the people with whom I had brainstorming discussions about my work: Marco Lavalle, Andreas Reigber, Eric Pottier, Maxim Neumann and Laurent Ferro-Famil. I am also infinitely grateful for the help of Juan Manuel Sanchez Lopez and Rafael Schneider who gave up their time to help with the proof-reading of my thesis.

I am immensely grateful for the sponsorships provided to me for my PhD, which includes many people, however, in brief the list most certainly has to include Nick Walker from eOsphere Ltd, Dr. Iain Anderson from Defence Science and Technology Laboratory of the Ministry of Defence UK, Tony Kinghorn and Neil Whitehall from the Electro Magnetic Remote Sensing Defence Technology Centre and Ralf Horn and Matteo Nannini from the German Aerospace Agency (DLR).

Without their continuous adjustments on the directions of my research I would never have attained the results presented in this thesis.

I would also like to thank the people within DLR who first planted the seed for research within me (or perhaps first watered it). Without them I surely would not have started my PhD in Edinburgh or developed such passion for research. To name but a few of these people, I have picked a small sample: Irena Hajnsek, Florian Kugler, Luca Marotti, Kostas Papathanassiou and Rafael Schneider.

I leave the last paragraph for a special thank to everybody who has been close to me during my PhD, especially in the beginning when home looked so far away. Amongst these people are my parents, Franco and Anna who accepted the idea to live afar from their son and provide me with an opportunity of attempting my aspirations. My friends amongst others, Pasquale Bellotti, Giampaolo Cesareo, Vincenzo Costa and Giuliano Raimondo. Last but not least I would like to thank my girlfriend, Greer Gardner, who supported me in all my decisions putting my aspirations before everything.

Contents

Acronyms

EM	Electro magnetic field
SAR	Synthetic aperture radar
PF	Polarisation fork
FOLPEN	Foliage penetration
POLSAR	Polarimetric SAR
POLinSAR	Polarimetric SAR interferometry
X-pol	Cross-polarisation
Co-pol	Co-polarisation
CR	Corner reflector
RedR	Reduction Ratio
SCR	Signal to clutter ratio
SNR	Signal to noise ratio
CNR	Clutter to noise ratio
pdf	Probability density function
CDF	Cumulative distribution function
DF	Discrete probability function
ROC	Receiver operating characteristic
H	Horizontal
V	Vertical
SLC	Single look comples image
DEM	Digital elevation model
LOS	Line of sight
RVoG	Random volume over ground
OVoG	Oriented volume over ground
RCS	Radar cross section
STD	Single target detector
PTD	Partial target detector
hfs	Historical fire scar

Symbols

X_1, X_2	X-Pol Nulls
C_1, C_2	Co-Pol Nulls
S_1, S_2	X-Pol Max
γ	Complex polarimetric coherence
γ_d	Detector
$\underline{k}$	Scattering vector
k_1, k_2 and k_3	Components of the scattering vector $\underline{k}$
$\underline{\omega}$	Scattering mechanism
$\underline{\omega}_T$	Target of interest (scattering mechanism)
$\underline{\omega}_P$	Perturbed-target (scattering mechanism)
$i(.)$	Complex image
$[S]$	Scattering (Sinclair) matrix
$[C]$	Covariance matrix
$[T]$	Coherency matrix
$[U]$	Unitary rotation matrix
$\langle . \rangle$	Finite averaging
$E[.]$	Expected value (infinite averaging)
$\underline{k}^T$	Transpose of $\underline{k}$
$\underline{k}^*$	complex conjugate of $\underline{k}$
$\|\underline{k}\|$	Modulus of $\underline{k}$
$|\underline{k}_i|$	Amplitude of the component $\underline{k}_i$
$\left(\frac{|b|}{|a|}\right)^2, \left(\frac{|c|}{|a|}\right)^2$	Reduction ratios $RedR$
T	Detector threshold
P_T	Power of target component
P_C	Power of clutter component
$[I]$	Identity matrix
$[A]$	Weighting matrix
$N(.)$	Gaussian (Normal) distribution

$\Gamma(.)$	Gamma distribution
$\delta(.)$	Dirac function
X_i	Random variable
x_i	Realisation of the random variable X_i
P_D	Probability of detection
P_F	Probability of false alarm
P_M	Probability of missed detection

Chapter 1
Introduction

Surveillance is the monitoring of human activities (or other changing information), usually accomplished for military, law enforcement or security reasons. Even though the principal interest of surveillance is dedicated to military activities, many vital civilian applications (i.e. security) can be listed. Clearly, from the military point of view, information about the enemy location has remarkable tactical advantages influencing the outcome of a battle. Regarding the benefits for law enforcement and security, monitoring illegal activities is essential to prevent crime. For instance, the monitoring of costal areas (or generally water regions) remains one of the most complex issues. A few examples of illegal activities can be illegal fishing (which impoverish the marine flora) or the transport of contraband. Mountainous, forested or desert areas often constitute entrance points for contraband and forested areas offer protection for hunted criminals or illegal logging (Illegal-Logging.info, Leipnik and Donald 2003). In order to follow the fast growth of modern technology, surveillance needs to exploit updated methodologies. The entire thesis is based on this concept: the development of a new system able to perform surveillance on large scale with the minimum effort and the maximum performance possible.

One of the first challenges of surveillance is the need to cover vast areas in a short time. In this context, the capability to monitor from a distance (i.e. remote sensing) constitutes a large strategic advantage (Campbel 2007). The latter becomes particularly vital when the surveillance by continuous in situ inspections is unfeasible (e.g. oceans, deserts or forests). In many cases, the only way to approach the regions under examination is by an aircraft. Moreover, a visual inspection is practicable only with solar illumination and favourable weather conditions, reducing strongly the reliability of such solutions. Additionally, the single coverage of one flight is relatively small and the aircraft has to perform several flights over the same region in order to cover an area large enough. On the other hand, a satellite will naturally come back periodically (depending on its orbit) to the same area and the coverage of the single acquisition is a large strip

A. Marino, *A New Target Detector Based on Geometrical Perturbation Filters for Polarimetric Synthetic Aperture Radar (POL-SAR)*, Springer Theses, DOI: 10.1007/978-3-642-27163-2_1, © Springer-Verlag Berlin Heidelberg 2012

with width that goes from tens to hundreds of kilometres (the length can be longer) (Chuvieco and Huete 2009). Currently, the repeat pass time of a single satellite (here we refer to the radar one, since these will be employed in the validation) is still too long for some applications, however if data from several satellite platforms (or using different look angles) are acquired, the temporal distance can be reduced to less than one day. Recently, several projects have been involved in designing a constellation of radar satellites, as will be explained in the following.

We can conclude that remote sensing is indispensable for large scale coverage. Now, we are concerned with the system (or sensor) which best suits our detection requirements. In this thesis we decided to use Synthetic Aperture Radar (SAR). SAR is an active remote sensing system exploiting the electromagnetic (EM) field in the microwave region (Rothwell and Cloud 2001; Richards 2009). Compared with less sophisticated radar sensors (i.e. scatterometers), the SAR architecture has the remarkable advantage of achieving very high sptial resolution reflectivity images of the observed scene. As a consequence, it provides enhanced discrimination between targets falling in different pixels. After about 30 years of continuous refinements, SAR is nowadays widely utilised with several airborne and satellite platforms exclusively dedicated to SAR sensors. In particular, a large number of radar satellites have been providing SAR images for several decades (i.e. ERS-1, ERS-2, ASAR-LANDSAT, RADARSAT, SIR-X, SIR-C, etc.). With the advancement of hardware, the latest satellites can provide data with improved radiometric characteristics and perform polarimetric acquisitions. Examples are ALOS-PALSAR (from JAXA: Japanese Space Exploration Agency) (ALOS 2007), RADARSAT-2 (from the Canadian Space Agency) (Slade 2009) and TERRASAR-X (from DLR: German Aerospace Agency) (Fritz and Eineder 2009; Lee and Pottier 2009). Recently, several satellite constellations have been designed, a few examples of which are already launched or being developed: CosmoSkyMed (Agenzia Spaziale Italiana 2007) composed of 4 X-band satellites and the planned Sentinel-1 (ESA, European Space Agency) (Attema et al. 2007). Considering the enormous costs of constructing a satellite and sending it into orbit, the large presence of sensors dedicated to SAR is a clear indicator of the significant contribution of radar in remote sensing.

For many applications, the direct competitors of SAR are optical sensors (Campbel 2007). In the basic principles, the two technologies reveals several similarities since both acquire the electromagnetic wave scattered from objects, although most of the optical systems are passive (i.e. they do not transmit but only receive the EM field). Except the passive architecture, the main difference consists in the exploitation of different frequencies (or wavelengths). The scattering from an illuminated object behaves dissimilarly depending on the wavelength (Cloude 1995, Rothwell and Cloud 2001; Woodhouse 2006). As a consequence, this diversity demarks the areas of applicability of the two techniques. In order to appreciate the advantages of one system over the other, the interaction between the EM wave and matter must be understood. In general, an EM field interacts with objects having dimensions similar or bigger than the wavelength (Stratton 1941; Rothwell and Cloud 2001; Cloude 1995; Woodhouse 2006). Consequently, small objects become rather transparent allowing the EM wave to pass through some cluster media with little attenuation. For instance, clouds are reasonably transparent at microwaves (especially if low frequencies are considered). Therefore, measurements can be performed under almost any weather condition (Richards 2009; Woodhouse 2006). In some contexts, end users judge this property as the most significant benefit of SAR. However, many other advantages of comparable importance can be listed. For instance, a forest canopy can be modelled as a cluster medium penetrable to some extent by the EM waves (especially where wavelengths are greater than about 20 cm). Going through tens of metres, the EM field collects information about inner forest layers. This makes radar particularly suited for the study of vegetation (Campbel 2007; Woodhouse 2006; Treuhaft and Siqueria 2000; Cloude et al. 2004).

Another noteworthy benefit of microwaves is the possibility of measuring the phase of the EM field (i.e. coherent acquisition). With the purpose of explaining the importance of the phase information, two of the most powerful methodologies in radar remote sensing can be listed: interferometry and polarimetry (Bamler and Hartl 1998; Cloude 2009; Papathanassiou and Cloude 2001). Please note, optical systems can be polarimetric even though they do not acquire the phase. However, they generally collect only two polarisations (co- and cross-polar). As will be explained in the following, in this way they do not extract the entire polarimetric information of the observed target. Without phase measurements, the number of acquisitions needed to have a complete polarimetric picture is generally too large and the actual sensors become significantly more complex (Cloude 2009).

In conclusion, we believe that SAR meets the reliability requirements of any weather and beneath-canopy detection. Considering the intrinsic capabilities of polarimetry to discriminate between several targets, we decided to exploit it in the development of a new detector algorithm. Polarimetry studies the geometrical properties of the EM wave propagating in space (Cloude 2009; Lee and Pottier 2009; Mott 2007; Ulaby and Elachi 1990; Zebker and Van Zyl 1991). Specifically, it is related to the shape that the electric field draws on the plane transverse to the direction of propagation. Interestingly, when two different targets are illuminated by the same polarised radiation, they are likely to scatter a wave with different polarisation states. Therefore, the polarisation state can be exploited to discriminate among observed targets. For instance, a metallic wire scatters a field linearly oriented in the same orientation as the wire. As a general rule, two objects are expected to have different polarimetric behaviour if their shape (considering the parts with dimensions comparable to, or bigger than, the wavelength) and material are different. The direct consequence is the possibility to describe a target using its polarimetric behaviour. A correspondence can be arranged between real and polarimetric targets exploiting vectors in a defined algebraic space (Cloude 1995; Huynen 1970; Kennaugh and Sloan 1952). The complete characterisation of a polarimetric target requires at least four acquisitions (which can be reduced to 3 under certain conditions which are common for Earth observation). Such datasets are named "quad-polarimetric" or "fully polarimetric". When only two polarisations are acquired the dataset is regarded as "dual-polarimetric". The reason for acquiring only two polarisations is mainly linked with the availability of simpler (lower cost) hardware, the request of higher resolution or the need for a smaller amount of data to store (or transmit). Nowadays, virtually all radar sensors (including satellites) are able to acquire quad-polarimetric data, since it was demonstrated that several applications can be achieved exclusively by the use of polarimetry (Lee and Pottier 2009).

Summarising, microwave remote sensing has several advantages compared with other systems. Specifically, its penetration capability makes it very suitable for any weather measurements (and under canopy detection), while the coherent acquisition allows the exploitation of polarimetry which is of remarkable value for the recognition of specific targets. For these reasons, we believe that SAR polarimetry is undeniably suited to achieve the goal of this thesis, which is target detection in any weather condition and under canopy cover, for surveillance purposes.

The clear strategic advantage of radar polarimetry has led to the development of many detectors and classifiers, which at a general level can be classed into those that are physically based or those that are statistically based (Chaney et al. 1990; Novak et al. 1993). An introduction will be provided in the following chapters, but here we just mention the main differences. The physical approach performs the detection by exploiting the particular polarimetric signature of the target. This signature is narrowly related to the electromagnetic interactions of the scatterers and can allow the retrieval of physical parameters. The statistical techniques use the information kept in the stochastic nature of the scattering. The drawback of statistical techniques is that the physics behind the process is often lost and the retrieval of parameters starting from the statistics is particularly challenging.

The evaluation of the performance of a new algorithm is always complex and several strategies can be followed. In our treatment, we want to propose a simple series of criteria to be fulfilled. They are essentially based on two probabilities:

(1) Low probability of missing a target on the scene (i.e. missed detection). Specifically, two target typologies are exceptionally relevant, since their detection is renowned to be particularly difficult:

  1.1. **Targets under foliage cover** (Fleischman et al. 1996). This is noteworthy for two main reasons: firstly, forests can be shelter for illegal activities and secondly, because patrolling forested areas with ground inspections is highly complicated. To this we can add that the inspection is in many cases impracticable with optical systems because the tree canopy can represent a barrier for direct optical detection.

  1.2. **Small targets**. Most detectors are based on the target brightness (i.e. the amount of backscattering), consequently small or dielectric targets are easily lost in the image. Our concern is to develop an algorithm able to detect this target typology with a performance comparable to bright or big targets.

(2) Low probability of positive detection in absence of an actual target (i.e. false alarm). False alarms are particularly troublesome since they can trigger alert messages when the actual target is absent. This can lead to expensive visits and mistrust regarding the system. Specifically, we would like to meet two requirements:

    2.1. **Statistical stability**. Any algorithm working on real data can be interpreted as a statistical entity, since the observables can generally be modelled as random variables (in the following, this concept will be explained in more detail). Therefore, the algorithm must be robust and stable from the statistical point of view, providing small theoretical probabilities to return false positive (Kay 1998).

    2.2. **Robustness against bright natural targets**. As mentioned previously, most detectors are based on evaluating the points brightness. However, in a SAR scene many bright targets are merely the result of geometrical effects (i.e. layover), and they do not constitute authentic targets (i.e. they are false alarms). The new algorithm must be able to deal with this typology of points, rejecting them from the detection.

In the following, the content of the chapters is illustrated:

(1) Chapter two will introduce Synthetic Aperture Radar (SAR), providing the basic tools and fundamental knowledge for the development of the detector. For the sake of brevity, the formulation is kept brief and we decided to introduce only concepts which have a direct utilisation in the proposed detector.

(2) The third chapter is dedicated to radar polarimetry introducing the concepts of wave and target polarimetry. This theoretical chapter is particularly significant for the purpose of the thesis. Specifically, a fundamental differentiation is drawn between single (deterministic or coherent) and partial (statistical) targets.

(3) The fourth chapter presents the development of the mathematical formulation of the detector. Two different approaches are followed: a physical and geometrical one, in order to provide a larger picture of the algorithm. The optimisation of the detector parameters, interpreted as a mathematical entity, are considered.

(4) The fifth chapter examines the proposed detector as a statistical entity performing the optimisations from a statistical point of view. In actual fact, a statistical approach is indispensable if we want to characterise quantitatively the performances of the detector. Please note, however, that the evaluation of the statistics does not make the final algorithm a statistical detector, since it is still intimately based on the scattering physics. In particular, the probability density function (pdf) of the detector is analytically derived.

(5) The sixth chapter concerns the validation of the detector with real data. In this chapter, airborne data (E-SAR, DLR) is utilised since they represent a best

scenario for the detector providing high resolution and signal to noise ratio (SNR). A comparison with other polarimetric detectors is also provided.

(6) The last chapter treats the validation of the detector with satellite data (ALOS-PALSAR, RADARSAT-2 and TERRASAR-X). They represent a harder scenario for a series of reasons, but they are particularly advantageous for coverage purposes.

References

AGENZIA SPAZIALE ITALIANA (2007) COSMO-SkyMed SAR Products Handbook

ALOS (2007) Information on PALSAR product for ADEN users

Attema E, Bargellini P, Edwards P, Levrini G, Lokas S, Moeller L, Rosich-tell B, Secchi P, Torres R, Davidson M, Snoeij P (2007) Sentinel-1, The Radar Mission for GMES, Operational Land and Sea Services. esa bulletin 131

Bamler R, Hartl P (1998) Synthetic aperture radar interferometry. Inverse Problems 14:1–54

Campbel JB (2007) Introduction to remote sensing. The Guilford Press, New York

Chaney RD, Bud MC, Novak LM (1990) On the performance of polarimetric target detection algorithms. Aerospace Electron Syst Mag IEEE 5:10–15

Chuvieco E, Huete A (2009) Fundamentals of satellite remote sensing. Taylor & Francis Ltd, London

Cloude RS (1995) An introduction to wave propagation & antennas. UCL Press, London

Cloude SR (2009) Polarisation: applications in remote sensing. Oxford University Press, 978-0-19-956973-1

Cloude SR, Corr DG, Williams ML (2004) Target detection beneath foliage using polarimetric synthetic aperture radar interferometry. Waves Random Complex Media 14:393–414

Fleischman JG, Ayasli S, Adams EM (1996) Foliage attenuation and backscatter analysis of SAR imagery. IEEE Trans Aerospace Electron Syst 32:135–144

Fritz T, Eineder M (2009) TerraSAR-X, ground segment, basic product specification document. DLR. Cluster Applied Remote Sensing

Huynen JR (1970) Phenomenological theory of radar targets. Technical University The Netherlands, Delft

Kay SM (1998) Fundamentals of statistical signal processing, Volume 2: detection theory. Prentice Hall, Upper Saddle River

Kennaugh EM, Sloan RW (1952) Effects of type of polarization on echo characteristics. In: Ohio State University Research Foundation Columbus Antenna Lab (ed)

Lee JS, Pottier E (2009) Polarimetric radar imaging: from basics to applications. CRC Press, Boca Raton

Leipnik MR, Donald PA (2003) GIS in law enforcement: implementation issues and case studies (International forensic science and investigation). Taylor & Francis, London

Mott H (2007) Remote sensing with polarimetric radar. Wiley, Hoboken

Novak LM, Burl MC, Irving MW (1993) Optimal polarimetric processing for enhanced target detection. IEEE Trans Aerospace Electr Syst 20:234–244

Papathanassiou KP, Cloude SR (2001) Single-baseline polarimetric sar interferometry. IEEE Trans Geosci Remote Sens 39:2352–2363

Richards JA (2009) Remote sensing with imaging radar—signals and communication technology. Springer, Berlin

Rothwell EJ, Cloud MJ (2001) Electromagnetics. CRC Press, Boca Raton

Slade B (2009) RADARSAT-2 product description. Dettwiler and Associates, MacDonals

Stratton JA (1941) Electromagnetic theory. McGraw-Hill, New York

Treuhaft RN, Siqueria P (2000) Vertical structure of vegetated land surfaces from interferometric and polarimetric radar. Radio Sci 35:141–177

Ulaby FT, Elachi C (1990) Radar polarimetry for geo-science applications. Artech House, Norwood

Woodhouse IH (2006) Introduction to microwave remote sensing. CRC Press Taylor & Frencies Group, Boca Raton

Zebker HA, Van Zyl JJ (1991) Imaging radar polarimetry: a review. Proc IEEE, 79

Chapter 2
Synthetic Aperture Radar

2.1 Radar Remote Sensing with SAR

Radar is an active microwave remote sensing system, first developed during the Second World War with the purpose of evaluating distances between targets (aircrafts, ships, etc.) and the antenna used to send and receive an Electromagnetic (EM) pulse (Woodhouse 2006; Brown 1999). After the war, the technique stopped being exploited exclusively for aircrafts/ships ranging and found interesting applications in remote sensing of the environment as well. Since its introduction in the remote sensing scientific community, radar has experienced a rapid growth, with the proliferation of numerous applications/techniques exploiting different features of the coherent acquisition of microwaves (Woodhouse 2006).

Microwave has some similarity with optical remote sensing since both acquire the electromagnetic wave scattered from objects on the scene (the similarity is even closer with a LIDAR system). However, the main difference is related to the use of a longer wavelength (i.e. lower frequency), which at the same time represents the foremost radar advantage (Richards 2009). A longer wavelength allows the coherent acquisition of the EM field (i.e. acquisition of amplitude and phase). The information associated with the phase can be exploited with techniques like interferometry and polarimetry which cannot be easily obtained with optical systems (here stereoscopy is not considered as on interferometric technique since it does not work with interferometric fringes) (Bamler and Hartl 1998; Cloude 2009; Papathanassiou and Cloude 2001). In general, the EM radiation interacts with objects with similar or bigger dimension than the wavelength (Stratton 1941; Rothwell and Cloud 2001; Cloude 1995; Woodhouse 2006). Consequently, objects that are small (compared with the wavelength) appear rather transparent to the radiation, and the wave is able to penetrate cluster mediums, composed by collections of particles. For instance, clouds are reasonably transparent to microwaves (especially in lower SAR frequencies) providing measurements with almost

A. Marino, *A New Target Detector Based on Geometrical Perturbation Filters for Polarimetric Synthetic Aperture Radar (POL-SAR)*, Springer Theses, DOI: 10.1007/978-3-642-27163-2_2, © Springer-Verlag Berlin Heidelberg 2012

any weather conditions. The forest canopy is another example of medium penetrability to some extent by the EM wave. This is one of the major advantages of surveying vegetation with radar. Due to the penetration (which can be tens of meters), the radiation collects information about the forest inner layers (Campbel 2007; Woodhouse 2006; Treuhaft and Siqueria 2000; Cloude et al. 2004).

The Synthetic Aperture Radar (SAR) is an ingenuous radar system which can acquire data with very high resolution. In a standard monostatic architecture, the system is composed of a platform (i.e. airborne or satellite) with the same antenna for transmitter and receiver (Franceschetti and Lanari 1999; Curlander and McDonough 1991; Massonnet and Souyris 2008). While the platform passes over the scene, the antenna transmits a series of EM pulses. Once the pulse reaches an object, an electrical current is exited over the object surface and (generally) generates an EM wave scattered back. Part of the radiation backscattered is recollected by the antenna on the platform (Fig. 2.1). Clearly, different arrangements can be considered, such as a bistatic SAR, where two different antennas are utilised for transmitter and receiver and they generally fly on two different platforms (Cherniakov 2008; Willis 2005). In this thesis, the focus is on monostatic sensors, although the proposed detector can be generalised to bistatic systems (as shown in the following chapters).

The platform moves along the *azimuth* direction with the antenna generally focused on a direction orthogonal to the azimuth: *range* (or slant-range). If the direction of observation is along the platform *nadir* (i.e. straight below the platform) the system is defined *boresight*. On the other hand, when the direction is inclined with an angle ϑ from the zenith, it is defined *side-looking* (the angle ϑ is called *look* angle). A side-looking solution is conventionally to be preferred to boresight for the rejection of range ambiguities (as will be shown in the following) (Franceschetti and Lanari 1999).

The acquisition process is achieved by transmitting a radio pulse (i.e. narrowband signal) and receiving the EM wave backscattered by the targets on the scene. In a classical radar system, the time delay from transmission to reception is related to the speed of the propagating wave and the distance from antenna to object:

$$\Delta t = \frac{2r}{c}, \qquad (2.1)$$

where r is the distance between sensor and scatterer and c is the speed of light.

In this basic arrangement, the resolution in range depends on the length of the pulse. Two scatterers can be separated if their distance is bigger than half the duration of the pulse, otherwise the two pulses will overlap each other. Hence, if we define with τ the temporal length of the pulse, the resolution will be:

$$\delta_{sr} = \frac{c\tau}{2} = \frac{c}{2W}, \qquad (2.2)$$

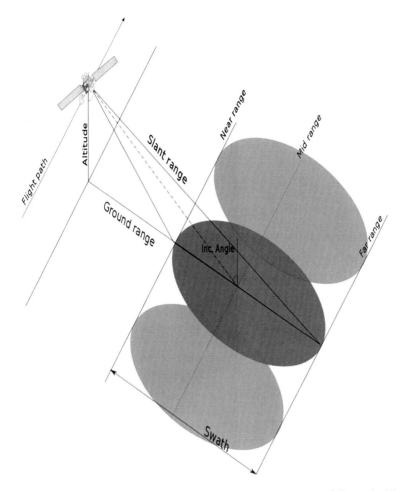

Fig. 2.1 SAR acquisition geometry for a monostatic system (courtesy of Fernando Vicente Guijalba)

where W is the bandwidth of the pulse. Therefore, in order to achieve high resolution the bandwidth must increase, leading to very short effective pulses which are generally not realisable in the designed bandwidth of the system. With the intention of achieving high resolution without decreasing the pulse duration, a frequency modulation was introduced (Curlander and McDonough 1991). The obtained pulse is called a *chirp* and is a linear frequency modulation of the narrowband pulse. It can be written as:

$$f(t) = \cos\left(\omega t + \frac{\alpha t^2}{2}\right) rect\left[\frac{t}{\tau}\right], \tag{2.3}$$

where $\omega = 2\pi f$ is the angular frequency and f is the carrier frequency, *rect* is the rectangular function of duration τ and α is the chirp rate related with the bandwidth W as $\alpha\tau = 2\pi W$. With the chirp, the bandwidth can be increased without reducing the duration of the pulse τ. In order to retrieve the actual scene information, the return must be cleaned from the alteration introduced by the linear phase modulation. This can be accomplished with a match filter with the chirp (and is known as range compression).

Regarding the azimuth resolution, the simplest system is a Real Aperture Radar (RAR). Here, all the points illuminated by the beam-width are collected together, hence they are inseparable. The resolution is dependent on the beam width (or aperture) of the antenna:

$$\Delta x_{RAR} = R\frac{\lambda}{L}, \tag{2.4}$$

where R is the distance between sensor and ground, λ is the wavelength exploited and L is the effective dimension of the antenna. In this configuration, the resolution depends on the distance to the sensor, making satellite applications suited only as scatterometers (Woodhouse 2006). In order to enhance the resolution the dimensions of the antenna or the frequency must increase. However, the frequency is fixed and the antenna cannot be excessively big for structural engineering reasons. A different solution had to be introduced.

The basic idea of the Synthetic Aperture Radar (SAR) is that a point on the ground is illuminated by the antenna not just with one single pulse but with a sequence of pulses. If all the acquisitions for the same point are collected, it will be similar to having performed a single acquisition with an antenna array with length (i.e. aperture) equal to the footprint X. After data compression the azimuth resolution becomes

$$\Delta x_{SAR} = \frac{L}{2}, \tag{2.5}$$

where L is the length of the antenna. Conversely to Eq. 2.4, the resolution improves when the effective dimension of the antenna L is reduced. This seems to contradict common sense since a smaller antenna has a larger beam width (hence a larger footprint). In actual fact, when L decreases, X increases and with it the synthetic antenna. As a consequence the array is larger and the final beam-width is sharper.

After the compression of the row data, the SAR image presents a map of the reflectivity (as a complex value) of the scene, where every pixel represents the coherent sum of the returns from the scatterers located in the resolution cell (Oliver and Quegan 1998). The reflectivity can be expressed with $\rho(r, x)$, where r and x are respectively the range and azimuth of the resolution cells. In any given pixel we have

$$\rho(r, x) = \sum_n \rho_n \delta(r - r_n, x - x_n), \tag{2.6}$$

Fig. 2.2 Estimation of
ground range resolution
(courtesy of Iain Woodhouse)

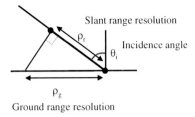

where δ is the Dirac function and r_n and x_n can move in the resolution cell considered. Therefore, the signal after the processing can be interpreted as a two dimensional complex signal (Massonnet and Souyris 2008).

2.2 Geometrical Distortions

A noteworthy divergence between (active) radar and (passive) optical systems is related to the acquisition arrangement. Radar was first designed with the purpose of acquiring distances between the sensor and the targets on the scene. This attribute is still central in the SAR acquisition strategy. The objects on the scene are arranged depending on the distances from the sensor rather than the location on the ground. Additionally, a radar system needs to be side-looking, where optical systems are often close to nadir. Due to this peculiar acquisition arrangement distortions are introduced on the reflectivity image and the latter cannot be compared straightforwardly with a map or photograph (Woodhouse 2006; Franceschetti and Lanari 1999; Campbel 2007).

Radar measures distances between sensor and scene, therefore distances on the ground (i.e. the horizontal plane where the scene lies) are not preserved and in *near-range* (region closer to the platform) the range resolution is larger than in *far-range* (region further from the platform). A new parameter can be introduced, regarded as *ground-range*, representing the distance measured along the projection of the range (now called *slant-range*) on the horizontal plane. Specifically, the ground range resolution can be calculated as

$$\Delta r_g = \frac{\Delta r}{\sin \vartheta_i}, \tag{2.7}$$

where ϑ_i is the local look angle. Figure 2.2 illustrates the concept of ground range resolution.

Equation 2.7 states that a boresight system (i.e. $\vartheta = 0$) has resolution equal to ∞, since in the hypothesis of plane wave any plane parallel to the ground surface lie only in one single resolution cell. Clearly, the plane wave hypothesis is unsuited in this case and a spherical wave must replace it. The final effect of the variability in ground range resolution is a non linear stretch along the range direction (the near range is compressed).

Fig. 2.3 Constant range and
constant Doppler curves. The
sensor moves along the x_g
axis (Mott 2007), as projected
onto the ground surface

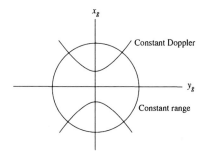

Equation 2.7 states the importance of side-looking architecture for image formation. Figure 2.3 shows the lines for equi-range and equi-Doppler (Mott 2007). The only way to avoid ambiguities between the two points above and below the y_g axis is simply to focus the antenna only on one side. Although indispensable for image formation, side looking is the cause of distortions in a SAR image.

Figure 2.4 presents the main distortions suffered by a radar image due to the side-looking architecture. The distortions are (Franceschetti and Lanari 1999):

a. Foreshortening: when a slope faces the sensor the illuminated area is compressed in less resolution cells. In other words a larger amount of ground lies in the same resolution cell since the apparent local look angle (calculated with the normal of the surface) is reduced. In a SAR image foreshortening produces a shift of the side of mountains (or generally slopes) facing the sensor in the direction of the sensor. Moreover, it is generally accompanied by a rising in backscattering since the number of scatterers in the same resolution cell increases (the energy of all the scatterers is compacted in a smaller area).
b. Layover: when the steepness of the slope facing the sensor is higher than the look angle the return from the top of the object comes before the one from the bottom. If compared with an optical image, the layover flips top and bottom. Layover is rather common in SAR imaging, since it affects all the vertical structures with changes in height bigger than the resolution cell (i.e. buildings, trees).
c. Shadowing: this effect is observable on the slopes opposite to the sensor and it can be interpreted as the opposite of foreshortening. The areas affected by shadowing are enlarged (along the range direction). When the slope is smaller than $\vartheta - \pi/2$, the areas become completely dark (in optical shadow).

In general, shadowing areas (i.e. slopes facing away from the sensor) are darker since the energy is spread over a larger area or they are not visible at all (additionally the surfaces scatter less as predicted by the Bragg model).

Another cause of distortion in a SAR image is associated with the dissimilar resolutions in range and azimuth. In the image formation, the range resolution depends on the bandwidth (Eq. 2.2) while the azimuth one depends on the length of the antenna (Eq. 2.5). As a result, the pixel will not generally be square but rather rectangular on the ground. A rectangular pixel stretches the image in the

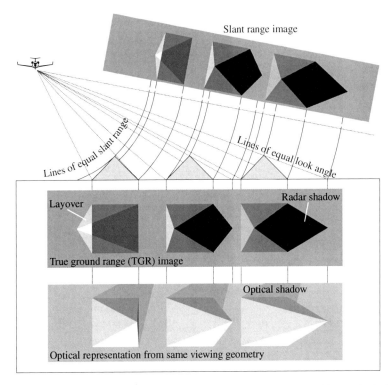

Fig. 2.4 Distortions suffered by the side looking architecture compared with a optical viewing geometry (courtesy of Iain Woodhouse)

direction where the resolution is higher (for satellite applications often this is the azimuth).

Due to the severe distortions affecting a SAR image, the latter cannot be overlapped straightforwardly with a map. As a first step, the image must be geocoded in order to correct the geometric distortions. Subsequently, it must be projected on a coordinate system with a geo-location. In the validation of the proposed detector, in the current work, geo-location of the images is not performed since the detector works with the physics of the backscattering (polarimetry) and this does not change with geo-location (as long as the geo-location is well defined) (Campbel 2007; Wise 2002).

2.3 Statistical Characterisation of Targets

The backscattered field acquired by a SAR system is a product of the interaction between objects on the scene and the microwave pulse sent by the transmitter antenna. The interaction is generally more consistent when the dimensions of the illuminated object are equal or bigger than the wavelength (Stratton 1941; Rothwell and Cloud 2001). In microwave remote sensing, the wavelength is around centimetres (X- or C-band) or tens of centimetres (S-, L- and P-band) while the resolution cell is around meters. For this reason, in the same resolution cell several scatterers contribute to the total backscattered field. For the theorem of superimposition of fields, all the EM waves coming from the same resolution cell are summed coherently together (generally the phases must be taken into account and the total power is not the sum of the power contributions) (Oliver and Quegan 1998; Rothwell and Cloud 2001).

If the return from the *ith* scatterer in the cell is $V_i e^{j\phi_i}$ the total return will be:

$$V_{re} + jV_{im} = V = \sum_{i=1}^{N} V_i e^{j\phi_i} = \sum_{i=1}^{N} V_i \cos \phi_i + j \sum_{i=1}^{N} V_i \sin \phi_i. \qquad (2.8)$$

The possibility to represent the EM wave with a complex number will be illustrated in the next section. Equation 2.8 describes the coherent sum of the contributions in the cell. If there is not a single dominant scatterer, the only way to extract information about the observed scatterers is to treat the problem with a statistical approach (Oliver and Quegan 1998). In fact, the number of observables (i.e. real and imaginary part of the total return) is smaller than the number of unknowns.

If N is big enough, we can apply the central limit theorem and say that the real and imaginary part of the return are normally distributed: $V_{re} \sim N(0, \sigma^2)$ and $V_{im} \sim N(0, \sigma^2)$. Their probability density functions (*pdf*) are

$$\begin{aligned} f_{V_{re}}(V_{re}) &= \frac{1}{\sqrt{2\pi\sigma^2}} \exp\left(-\frac{V_{re}^2}{2\sigma^2}\right), \\ f_{V_{im}}(V_{im}) &= \frac{1}{\sqrt{2\pi\sigma^2}} \exp\left(-\frac{V_{im}^2}{2\sigma^2}\right). \end{aligned} \qquad (2.9)$$

The mean is zero, $E[V_{re}] = E[V_{im}] = 0$, since the average of random real numbers with their sign is zero (Gray and Davisson 2004; Kay 1998; Papoulis 1965).

Furthermore, the real and imaginary parts are independent of each other which makes them uncorrelated:

$$E[V_{re} V_{im}] = E[V_{re}]E[V_{im}] = 0, \qquad (2.10)$$

where $E[.]$ stands for *expected value*.

The trend of a Gaussian random variable with zero mean and variable standard deviation σ is plotted in Fig. 2.5.

The SAR image displays the reflectivity of a scene and it can be represented as a matrix of complex numbers. The amplitude of such complex numbers keeps valuable information about the amount of backscattering coming from the

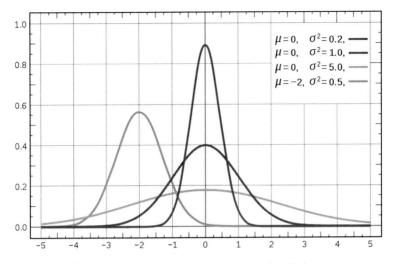

Fig. 2.5 Gaussian distribution with variable mean and standard deviation

resolution cell (since it is the square root of the power). Using the *pdf* of real and imaginary parts, it is possible to extract the joint *pdf* of amplitude and phase of the backscattering

$$f_{V\phi}(V, \phi) = \frac{V}{\sqrt{2\pi\sigma^2}} \exp\left(-\frac{V^2}{2\sigma^2}\right). \tag{2.11}$$

Integrating the expression in Eq. 2.11 on the entire interval of the phase, the *pdf* of the amplitude can be extracted

$$f_V(V) = \int_0^{2\pi} f_{V\phi}(V, \phi)d\phi = \frac{V}{\sigma^2} \exp\left(-\frac{V^2}{2\sigma^2}\right), \quad V \geq 0 \tag{2.12}$$

The latter corresponds to a Rayleigh distribution defined in $[0, \infty]$ and regarded as $V \sim Rayleigh(\sigma)$ (Papoulis 1965).

A quick way to characterise a random variable is using its principal modes. They can be obtained by integrating the expression of the *pdf* as shown in the following:

$$E[V] = \int_0^\infty V f_V(V)dV = \sqrt{\frac{\pi}{2}}\sigma, \tag{2.13}$$

$$E[V^2] = \int_0^\infty V^2 f_V(V)dV = 2\sigma^2, \tag{2.14}$$

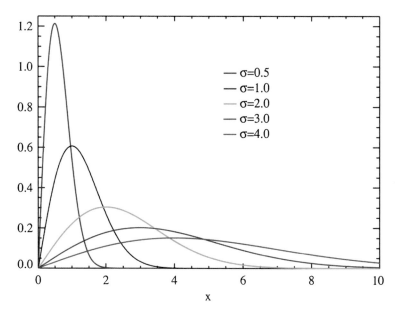

Fig. 2.6 Rayleigh distribution for variable σ

$$VAR[V] = E[V^2] - E[V]^2 = \frac{4 - \pi}{2}\sigma^2. \tag{2.15}$$

Figure 2.6 shows the Rayleigh distribution. As expected the probability of negative values is zero and the variation becomes bigger when the mean increases.

The *pdf* of the phase can be extracted as well and is uniformly distributed in $[0, 2\pi]$.

Once the statistical distribution of the amplitude is obtained, we can describe the power distribution of the backscattering, i.e. $W = V^2$. After some manipulation the *pdf* of the power is found as

$$f_W(W) = \frac{1}{2\sigma^2}\exp\left(-\frac{W}{2\sigma^2}\right), \quad W \geq 0. \tag{2.16}$$

which coincides with an exponential random variable. The latter is generally indicated with $W \sim Exp(\lambda)$, where λ is linked to the mean. The modes can be estimated:

$$E[W] = 2\sigma^2 = 1/\lambda, \tag{2.17}$$

$$E[W^2] = 8\sigma^4, \tag{2.18}$$

$$VAR[W] = E[W^2] - E[W]^2 = 4\sigma^4. \tag{2.19}$$

The mean of the exponential is ordinarily indicated with $1/\lambda$.

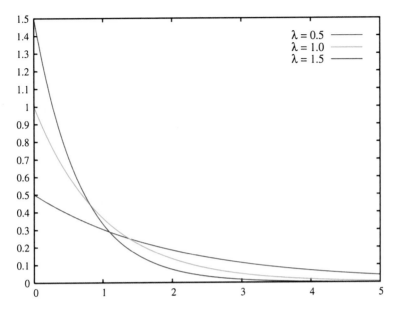

Fig. 2.7 Exponential distribution for variable λ

Figure 2.7 presents the Exponential distribution when the mean is varied. As in the case of a Rayleigh distribution, the variability of the Exponential is large and the standard deviation increases linearly with the mean (they are actually the same). This huge variation can lead to significant estimation errors making the scattering description on the basis of a single pixel a challenge.

In general, to reduce the variability of random variables, the average of independent and identically distributed (*iid*) realisations can be considered (please note, not all random variables when averaged reduce their variability).

If $W_1, \ldots, W_N$ are Exponential distributions, then the variable $\gamma = \sum_{i=1}^{N} W_i$ is a Gamma distribution indicated as $\gamma \sim \Gamma(\vartheta, k)$ where k is the *shape* factor depending on the number of elements summed (i.e. $k = N$) while $\vartheta = 1/\lambda = 2\sigma^2$ is the *scale* factor depending on the mean of the Exponential variables. The *pdf* is equal to:

$$f_{\overline{W}}(\overline{W}) = \left(\frac{N}{2\sigma^2}\right)^N \frac{\overline{W}^{N-1}}{(N-1)!} \exp\left(-\frac{N\overline{W}}{2\sigma^2}\right). \tag{2.20}$$

And its modes are:

$$E[\overline{W}] = 2N\sigma^2, \tag{2.21}$$

$$VAR[\overline{W}] = 4\sigma^4. \tag{2.22}$$

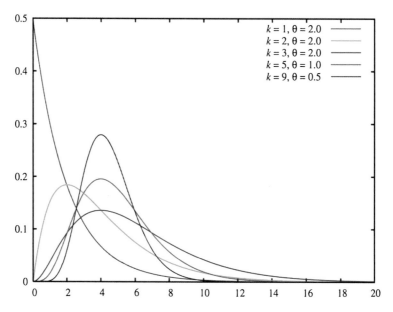

Fig. 2.8 Gamma distribution for variable k and ϑ

Commonly, the sum of Exponential is subsequently normalised by the number of samples N in order to achieve the average $\overline{W} = \dfrac{1}{N}\sum_{i=1}^{N} W_i$. The resulting random variable will be a scaled Γ distribution, $\overline{W} \sim \Gamma\left(\dfrac{\vartheta}{N}, k\right)$, and the modes will be:

$$E\left[\overline{W}\right] = 2\sigma^2, \tag{2.23}$$

$$VAR\left[\overline{W}\right] = \frac{4\sigma^4}{N}, \tag{2.24}$$

Figure 2.8 illustrates the Γ distribution with variable shape and scale factors.

The variance of the random variable is reduced by increasing the number of independent samples averaged. In order to obtain the desired reduction of variability, the sum must be performed on independent samples with the same mean (independent and identically distributed, *iid*). There are several methodologies to select independent samples in a SAR image. The most common (and the one used in this thesis) considers the average over neighbouring pixels with a moving window, but, more complicated strategies can be exploited.

2.4 Radar Cross Section

The aim of microwave remote sensing is extracting information from the EM wave scattered by an object. In particular, the power of the backscattering was the

Table 2.1 RCS for standard shapes (Franceschetti and Lanari 1999)

Shape	Sphere radius d	Square plate: side d	Triangular trihedral: side d	Square trihedral: side d
RCS $[m^2]$	πd^2	$4\pi\dfrac{d^4}{\lambda^2}$	$\dfrac{4\pi}{3}\dfrac{d^4}{\lambda^2}$	$12\pi\dfrac{d^4}{\lambda^2}$

subject of extensive studies when phase measurements where not yet feasible. A parameter called *radar cross section (RCS)* $\sigma[m^2]$ was introduced and it represents the area of an equivalent sphere (assumed as a perfect reflector) scattering the same amount of power as the target (Franceschetti and Lanari 1999; Woodhouse 2006). In a complex object, the power backscattered also depends on the angle of view of the target. Additionally, the induced currents on the surface of the object radiate in many directions and a directivity pattern of the object can be estimated.

If polar coordinates are introduced, the directions of incident and scattered wave can be characterised by the pairs (ϑ_i, φ_i) and (ϑ_s, φ_s) respectively. In conclusion, the *RCS* can be described as function of the direction for incident and scattered wave. Integrating over all the directions of the scattered wave (for a fixed incident wave), the total scattered power can be calculated as (Woodhouse 2006)

$$S(r, \vartheta_s, \varphi_s) = \frac{S_i(\vartheta_i, \varphi_i)\sigma(\vartheta_i, \varphi_i; \vartheta_s, \varphi_s)}{4\pi r^2}. \tag{2.25}$$

where r is the distance.

In the case of backscattering, the direction of incident and scattered waves is the same. Hence, $\vartheta_i = \vartheta_s$ and $\varphi_i = \varphi_s$.

For some simple shapes the calculation of the *RCS* is possible analytically (after various approximations). Some of these targets are considered in Table 2.1.

The *RCS* increases relatively fast in the case of corners (with the fourth power of the side), since they are able to collect the power of the illuminating wave in a narrow beam. The dependence on λ is related to the increased apparent dimensions of the surfaces.

The expression of the density of power can be used to estimate the power received by the system from an object at distance r:

$$P_r = \frac{PGA\sigma}{(4\pi)^2 r^4} = \frac{PG^2\lambda^2\sigma}{(4\pi)^3 r^4} = \frac{PA^2\sigma}{4\pi\lambda^2 r^4}. \tag{2.26}$$

where P is the peak power transmitted, G is the antenna gain and A is the antenna effective area.

It is interesting to note that the power goes down with the fourth power of the distance, which is because far from the source it propagates as a spherical wave with a dispersion of intensity as the square of the distance. Subsequently, the two way attenuation must be taken into account (by multiplying the two attenuations). The estimation of the theoretical power received is relevant in SAR image formation since a different power compensation for near and far range must be performed in order to have a reliable map of the scene reflectivity (Herwig 1992).

2.5 Polarimetric Acquisition: The Scattering Matrix

In this section the principles of radar polarimetric acquisition are introduced, specifically the formation of the scattering matrix, while a more complete treatment will be provided in the next chapter.

2.5.1 The Scattering Matrix

For the sake of brevity, the treatment will start from the definition of narrowband signals, leaving out the electromagnetism theory that deals with the derivation of the wave equations. If the bandwidth of the signal is small compared with the carrier frequency, the latter can be ignored and the electric (or magnetic) field can be represented with complex scalars. In the monochromatic case, the problem can be rigorously treated with fasors (Rothwell and Cloud 2001). Far from the source, the propagation is accomplished with a spherical wave that can be locally approximated as plane wave. The wave front is a plane and the electric and magnetic fields are orthogonal to the direction of propagation. Such propagation is equivalently regarded as Transverse ElectroMagnetic (*TEM*) since the fields lie in the transverse plane (Stratton 1941).

Figure 2.9 shows the coordinate system exploited.

The electric field can be written as

$$\underline{E} = E_x \underline{u}_x + E_y \underline{u}_y, \tag{2.27}$$

where the propagation is accomplished in the z direction and E_x, E_y, are complex numbers.

Hence, it can be written:

$$E_x = |E_x| e^{j\phi_x} \text{ and } E_y = |E_y| e^{j\phi_y}. \tag{2.28}$$

While the wave moves in space/time, the phase of the electric field changes. This effect can be taken into account with

$$E_x = |E_x| e^{j(\omega t - kz + \phi_x)} \text{ and } E_y = |E_y| e^{j(\omega t - kz + \phi_y)}. \tag{2.29}$$

where ω is the angular frequency $\omega = 2\pi f$, t is the time and k is the wavenumber $k = \dfrac{\omega}{c}$, with c speed of light in the medium considered. ϕ_x and ϕ_y give an initial phase for the two components.

Fig. 2.9 Coordinate system in agreement with the propagating wave (z direction)

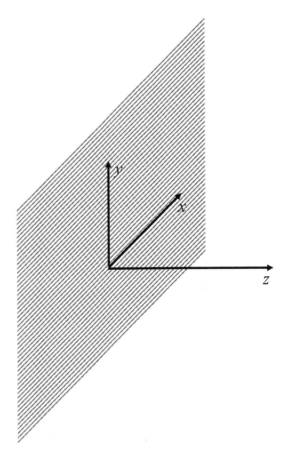

Once the information about the frequency is reintroduced in Eq. 2.27, the expression can be reconverted in time domain (real numbers) with:

$$e_x = \text{Re}\left\{|E_x|e^{j(\omega t - kz + \phi_x)}\right\} = |E_x|\cos(\omega t - kz + \phi_x),$$
$$e_y = \text{Re}\left\{|E_y|e^{j(\omega t - kz + \phi_y)}\right\} = |E_y|\cos(\omega t - kz + \phi_y). \tag{2.30}$$

The two components of the field (i.e. x and y) interact coherently with each other producing a resulting vector that moves on the plane of propagation. The polarisation of the EM field is related to the shape that the electric field draws on the transverse plane while the time passes. In this brief introduction, only stationary states of polarisation are considered. Specifically, if the electric field has a component only in one axis of the propagation plane its polarisation is defined to be linear (in general in order to have linear polarisations the two components must have the same phase).

When a target is excited by a wave which is linearly polarised in the x direction, $\underline{E}^i = E_x^i \underline{u}_x$, the scattered wave will be (Krogager 1993; Kennaugh and Sloan 1952; Mott 2007; Cloude 2009; Lee and Pottier 2009)

$$\underline{E}^s = E_x^s \underline{u}_x + E_y^s \underline{u}_y = \frac{1}{\sqrt{4\pi r^2}} \left[S_{11} E_x^i \underline{u}_x + S_{21} E_x^i \underline{u}_y \right]. \tag{2.31}$$

since the reradiated wave has generally a different polarisation from the incident one (as will be explained in the next section). Consequently, for any incident wave at least two measurements are necessary to characterise the scattered field (i.e. the x and y components).

The final requirement is to be able to describe the target scattering behaviour independently of the incident wave employed. Thus, the x component of the incident field alone is not sufficient since it is not sufficient to describe all the possible incident waves. The orthogonal component y must be considered as well. Therefore, a linear polarised field in the y direction can be transmitted and the return collected in the two components:

$$\underline{E}^s = E_x^s \underline{u}_x + E_y^s \underline{u}_y = \frac{1}{\sqrt{4\pi r^2}} \left[S_{12} E_y^i \underline{u}_x + S_{22} E_y^i \underline{u}_y \right]. \tag{2.32}$$

In summary, in order to describe completely the polarimetric behaviour of a target four acquisitions are needed: two to describe any scattered wave multiplied by two to describe any incident wave. The theorem of superposition of fields asserts that the four measurements can be done separately (but the target must not change). The four measurements can be collected in a matrix as

$$\underline{E}^s = \begin{bmatrix} E_x^s \\ E_y^s \end{bmatrix} = \begin{bmatrix} S_{11} & S_{12} \\ S_{21} & S_{22} \end{bmatrix} \begin{bmatrix} E_x^i \\ E_y^i \end{bmatrix} \frac{1}{\sqrt{2\pi r^2}}. \tag{2.33}$$

The matrix

$$[S] = \begin{bmatrix} S_{11} & S_{12} \\ S_{21} & S_{22} \end{bmatrix} \tag{2.34}$$

is called the *scattering* (or *Sinclair*) matrix. With the scattering matrix any stationary target illuminated by a wave with stationary polarisation can be completely characterised (Kennaugh and Sloan 1952). The hypothesis of stationarity seems to be unavoidable; however in the next section we will see that in the case of non-stationary processes we can still characterise a target exploiting its statistics.

When the scattering matrix is completely acquired in one single flight pass of the platform, the system is defined as *quad* polarimetric. The simultaneous acquisition is needed to reconstruct properly the polarimetric characteristic of the target, especially if this changes from one acquisition to another. However, in some cases, the sensor is not sufficiently complex to acquire [S] in one pass, but only half (for instance one column of the scattering matrix). In this scenario, the

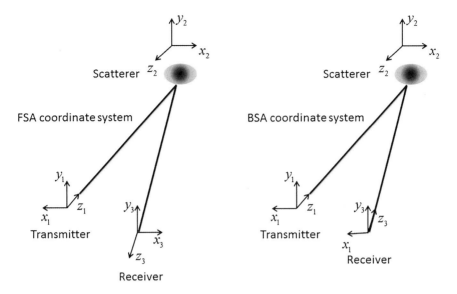

Fig. 2.10 Comparison of FSA and BSA coordinate systems

system is defined as *dual* polarimetric. Unfortunately, the latter is not able to describe completely a polarimetric target (Cloude 2009; Lee and Pottier 2009).

2.5.2 The Coordinate System

The correct selection of the coordinate system for a scattering problem is often a key point, since an advantageous selection can reveal symmetries which simplify drastically the treatment of the problem (Cloude 1995).

The most common choice is to set the coordinate system in agreement with the propagating wave (on the plane wave). This strategy takes the name of *Forward (anti-monostatic) Scattering Alignment (FSA)*, and it is probably the optimum alternative when the scattering occurs in any direction (as in the bistatic case). However, in general, the transmitter and receiver antennas are the same (i.e. monostatic system). In this situation, a coordinate system in agreement with the antenna can be employed since the antenna remains fixed. Such a coordinate system is regarded as *Back (Bistatic) Scattering Alignment (BSA)*. Figure 2.10 shows the comparison of the two arrangements (Boerner 2004).

The targets observed in a radar image are commonly reciprocal in the microwave range of frequencies. In the case of a monostatic arrangement and reciprocal medium the scattering matrix becomes symmetric since for the reciprocal theorem for antennas the same antenna behaves equally in transmission and reception (i.e. the scatterer can be interpreted as an antenna itself). Therefore, the two off-diagonal terms of [S] are the same. Please note, with the *FSA* arrangement the symmetry of [S] cannot be exploited. This symmetry introduces a significant simplification in the problem since only 3 complex numbers rather than 4 are necessary to characterise the target (Cloude 1995). Additionally, a symmetric matrix can be diagonalised with (generally) complex eigenvalues. The eigenvectors represent the optimum polarisations for the scattering problem, as will be presented in the next chapter (Huynen 1970; Kennaugh and Sloan 1952).

In this thesis, when it is not indicated otherwise, the *BSA* arrangement will be used, since it has been shown to be more advantageous for the study of backscattering problems.

References

Bamler R, Hartl P (1998) Synthetic aperture radar interferometry. Inverse Probl 14:1–54
Boerner WM (2004) Basics of radar polarimetry. RTO SET Lecture Series
Brown L (1999) A radar history of world war II, technical and military imperatives. Institute of Physics Publishing, Bristol and Philadelphia
Campbel JB (2007) Introduction to remote sensing. The Guilford Press, New York
Cherniakov M (2008) Bistatic radar: emerging technology. Wiley, Chichester
Cloude RS (1995) An introduction to wave propagation & antennas. UCL Press, London
Cloude SR (1995) Lie groups in EM wave propagation and scattering. Chapter 2 in electromagnetic symmetry. In: Baum C, Kritikos HN (eds). Taylor and Francis, Washington, ISBN 1-56032-321-3, pp 91–142
Cloude SR (2009) Polarisation: applications in remote sensing. Oxford University Press, Oxford, 978-0-19-956973-1
Cloude SR, Corr DG, Williams ML (2004) Target detection beneath foliage using polarimetric synthetic aperture radar interferometry. Waves Random Complex Media 14:393–414
Curlander JC, Mcdonough RN (1991) Synthetic aperture radar: systems and signal processing. Wiley, New York
Franceschetti G, Lanari R (1999) Synthetic aperture radar processing. CRC Press, Boca Raton
Gray RM, Davisson LD (2004) An introduction on statistical signal processing. Cambridge University Press, Cambridge
Herwig O (1992) Radiometric calibration of SAR systems. In: fundamentals and special problems of synthetic aperture radar (SAR), AGARD Lecture Series, Vol 182
Huynen JR (1970) Phenomenological theory of radar targets. Delft Technical University, The Netherlands
Kay SM (1998) Fundamentals of statistical signal processing, vol 2: Detection theory. Prentice Hall, Upper Saddle River
Kennaugh EM, Sloan RW (1952) Effects of type of polarization on echo characteristics. Ohio state University, Research Foundation Columbus, Quarterly progress reports (In lab)
Krogager E (1993) Aspects of polarimetric radar imaging. Technical University of Denmark, Lyngby
Lee JS, Pottier E (2009) Polarimetric radar imaging: from basics to applications. CRC Press, Boca Raton
Massonnet D, Souyris JC (2008) Imaging with synthetic aperture radar. EPFL Press, CRC Press, Boca Roton
Mott H (2007) Remote sensing with polarimetric radar. Wiley, Hoboken
Oliver C, Quegan S (1998) Understanding synthetic aperture radar images. Sci Tech Publishing, Inc., Raleigh

Papathanassiou KP, Cloude SR (2001) Single-baseline polarimetric SAR interferometry. IEEE Trans Geosci Remote Sens 39:2352–2363

Papoulis A (1965) Probability, random variables and stochastic processes. McGraw Hill, New York

Richards, JA (2009) Remote sensing with imaging radar—signals and communication technology. Springer-Verlag Berlin and Heidelberg GmbH & Co. KG, Germany

Rothwell EJ, Cloud MJ (2001) Electromagnetics. CRC Press, Boca Raton

Stratton JA (1941) Electromagnetic theory. McGraw-Hill, New York

Treuhaft RN, Siqueria P (2000) Vertical structure of vegetated land surfaces from interferometric and polarimetric radar. Radio Sci 35:141–177

Willis NJ (2005) Bistatic Radar. SciTech, Releigh

Wise S (2002) GIS basics. Taylor & Francis, London

Woodhouse IH (2006) Introduction to microwave remote sensing. CRC Press, Taylor & Francis Group, Boca Raton

Chapter 3
Radar Polarimetry

3.1 Introduction

The aim of this chapter is to provide the basic concepts and tools for the study of polarimetric observations. The literature in this context is vast (especially regarding the description of targets) and for the sake of brevity some issues are not covered. Instead, this chapter focuses on the tools actually utilised in the formulation of the polarimetric detector described in later chapters. For a thorough treatment of polarimetry the reader is directed to (Boerner 2004; Cloude 2009; Goldstein and Collett 2003; Mott 2007; Zebker and Van Zyl 1991; Lee and Pottier 2009).

3.2 Wave Polarimetry

This first section is focused on the polarisation of the plane wave, with no connection (apparently) with the physical target which has scattered the field. The next section will connect the results presented here to the physics of the scattering.

3.2.1 Polarisation Ellipse

An electromagnetic wave far from the source (generally $R_0 \geq 2\dfrac{D^2}{\lambda}$ where R_0 is the distance from the antenna, D is the aperture-width and λ is the wavelength) propagates as a locally plane wave (Stratton 1941; Cloude 1995a). If z is the direction of propagation, the electric field can be represented by

$$\underline{E} = \underline{u}_x E_x + \underline{u}_y E_y = |E_x|e^{j\phi_x}\left(\underline{u}_x + \underline{u}_y \frac{|E_y|}{|E_x|}e^{j(\phi_y - \phi_x)}\right), \qquad (3.1)$$

A. Marino, *A New Target Detector Based on Geometrical Perturbation Filters for Polarimetric Synthetic Aperture Radar (POL-SAR)*, Springer Theses, DOI: 10.1007/978-3-642-27163-2_3, © Springer-Verlag Berlin Heidelberg 2012

where $|E_x|$ and $|E_y|$ are the amplitudes of the electric field components and ϕ_x, ϕ_y its phases. Equation 3.1 states that in a plane wave the electric (and magnetic) field is orthogonal to the direction of propagation: *TEM* (Transverse electro magnetic) (Rothwell and Cloud 2001). The two vectors composing the electric field can be used to describe the polarisation state of the wave (Azzam and Bashara 1977; Goldstein and Collett 2003). A widely used parameter is the *normalised complex polarisation vector* $\underline{p}$ (Boerner 1981):

$$\underline{p} = \frac{\underline{E}}{|\underline{E}|}.$$ (3.2)

The ratio $\underline{p}$ is sufficient to describe completely the direction of the electric field on the plane transverse to the propagation direction (as long as the polarisation is stationary). The ratio $\underline{p}$ is a complex number, therefore the polarisation of the EM wave can be entirely characterised by two real parameters (i.e. real and imaginary part of $\underline{p}$). Equivalently the two Deschamps parameters α and ϕ (Deschamps 1951) can be exploited:

$$\phi = \phi_x - \phi_y \quad \text{and} \quad \frac{|E_y|}{|E_x|} = \tan(\alpha).$$ (3.3)

Please note that the Deschamps parameters are sufficient to characterise the polarisation of the EM field but not the total electric field since the absolute phase and amplitude are missed. However, the latter are not properties related to the wave polarisation but to the radar cross section and the distance of the target. Clearly, the two parameterisations describe the same physical entity hence they can be linked to each other by a relationship:

$$\rho = \left|\frac{E_y}{E_x}\right| e^{\phi} = \tan(\alpha) e^{\phi}.$$ (3.4)

The shape that the tip of the electric vector draws on the transverse plane is generally an ellipse and can be described by two angles and one amplitude. The angles are the *orientation* angle ψ defined in the interval $\psi \in \left[-\frac{\pi}{2}, \frac{\pi}{2}\right]$ and the *ellipticity* angle χ defined in $\chi \in \left[-\frac{\pi}{4}, \frac{\pi}{4}\right]$. Conventionally, positive and negative values of χ represents respectively left-handed (anti-clockwise) and right-handed (clockwise) rotations (Boerner 2004; Cloude 2009; Lee and Pottier 2009). Figure 3.1 depicts the polarisation ellipse in relation to the angles.

Again a unique link can be found between the ellipse angles and the polarisation ratio:

$$\rho = \frac{\cos 2\chi \sin 2\psi + j \sin 2\chi}{1 + \cos 2\chi \cos 2\psi}.$$ (3.5)

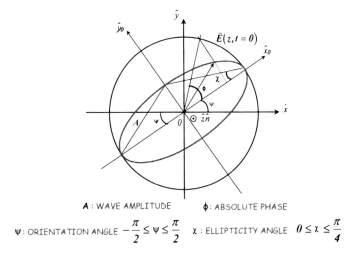

A : WAVE AMPLITUDE ϕ : ABSOLUTE PHASE

ψ : ORIENTATION ANGLE $\;-\dfrac{\pi}{2} \le \psi \le \dfrac{\pi}{2}\;$ χ : ELLIPTICITY ANGLE $\;0 \le \chi \le \dfrac{\pi}{4}$

Fig. 3.1 Polarisation ellipse (Boerner 2004) (Courtesy of Prof. E. Pottier and ESA)

While the link with the Deschamps parameters is

$$\cos 2\alpha = \cos 2\chi \cos 2\psi,$$

$$\tan \phi = \frac{\tan 2\chi}{\sin 2\psi}. \tag{3.6}$$

3.2.2 Jones Vectors

In the previous section the choice of parameters depends on the selected coordinate system (in our case horizontal and vertical). In order to generalise the treatment, the electric field must be expressed as the coherent superposition of two orthogonal components

$$\underline{E} = \underline{u}_m E_m + \underline{u}_n E_n, \tag{3.7}$$

where $\underline{u}_m$ and $\underline{u}_n$ are two generic orthogonal unitary vectors on the plane transverse to the propagation (Goldstein and Collett 2003; Beckmann 1968). The components E_m and E_n (complex numbers) can be used to define a vector, which is called the *Jones* vector:

$$\underline{E}_{mn} = \begin{bmatrix} E_m \\ E_n \end{bmatrix}. \tag{3.8}$$

The Jones vector is a two dimensional complex vector, therefore it has four degrees of freedom. In order to keep the formulation general, a procedure is needed

Table 3.1 Expression of the complex polarisation ratio ρ for different bases

Polarisation	ψ	χ	ρ_{HV}	$\rho_{45°135°}$	ρ_{LR}
Linear horizontal	0	0	0	-1	1
Linear vertical	0	$\dfrac{\pi}{2}$	∞	1	-1
45° linear	0	$\dfrac{\pi}{4}$	1	0	j
135° linear	0	$-\dfrac{\pi}{4}$	-1	∞	$-j$
Left handed circular	$\dfrac{\pi}{4}$	Any	j	j	0
Right handed circular	$-\dfrac{\pi}{4}$	Any	$-j$	$-j$	∞

to modify the basis starting from a generic Jones vector. If another basis $\underline{u}_i$ and $\underline{u}_j$ is selected the vector becomes

$$\underline{E}_{ij} = \begin{bmatrix} E_i \\ E_j \end{bmatrix}. \tag{3.9}$$

The transformation able to map the vector in the new basis is performed with a unitary 2×2 matrix

$$E_{ij} = [U_2] E_{mn}. \tag{3.10}$$

The transformation needs to be unitary since a rotation of the axis does not change the vector length (it can be interpreted as a conservation law) (Cloude 1995c; Bebbington 1992).

The link of $[U_2]$ with the complex polarisation ratio of the initial basis is

$$[U_2] = \frac{1}{\sqrt{1 + \rho\rho^*}} \begin{bmatrix} 1 & -\rho^* \\ \rho & 1 \end{bmatrix} \begin{bmatrix} e^{\phi_i} & 0 \\ 0 & e^{\phi_i} \end{bmatrix}, \tag{3.11}$$

where ϕ_i is a phase factor. Remarkably, this phase (in the following defined as *absolute* phase) can be neglected in the case of single pass polarimetry although it keeps information about the observed target (Cloude and Papathanassiou 1998; Papathanassiou 1999). In conclusion, the complex polarisation ratio ρ is dependent on the basis considered. Table 3.1 presents the values of ρ for frequently utilised bases.

3.2.3 Stokes Vectors

In the previous section, the treatment of wave polarimetry assumes an implicit hypothesis: stationarity in time. Obviously, the electric field is not stationary since it keeps oscillating, tracing the polarisation ellipse. It is the later which remains stationary (i.e. the same in time). This does not represent the general scenario where the polarisation ellipse is a function of time and changes shape. The aim of this section is to introduce the mathematical tools required to describe a non-stationary wave polarisation (Beckmann 1968).

If the EM wave changes its polarisation in time, it will be regarded as partially polarised, in contraposition with the completely or pure polarised one (treated in the previous section). When the polarisation changes in time, instantaneous observations become insufficient to characterise completely the wave, therefore averaged information are required. The Jones vector can be used to calculate means and cross-correlation of the two components of the field, providing the statistics characterisation of the moments. A wave coherence matrix (or Wolf matrix) is defined as

$$[J] = \left\langle \underline{E}\,\underline{E}^{*T} \right\rangle = \begin{bmatrix} \langle E_H E_H^* \rangle & \langle E_H E_V^* \rangle \\ \langle E_V E_H^* \rangle & \langle E_V E_V^* \rangle \end{bmatrix} = \begin{bmatrix} J_{HH} & J_{HV} \\ J_{VH} & J_{VV} \end{bmatrix}, \tag{3.12}$$

where $\langle . \rangle$ represents the temporal/ensemble averaging (Jones 1941; Wolf 2003). $[J]$ is positive definite and has Hermitian symmetry. The diagonal terms correspond to the components power, hence the sum of the diagonal terms, i.e. $Trace\{[J]\}$ is the power of the wave. On the other hand, the diagonal terms are the cross-correlations between components. If there is no correlation (i.e. $J_{HV} = J_{VH} = 0$) the wave is completely unpolarised and the power is distributed equally in any two orthogonal axes (in particular $J_{HH} = J_{VV}$) (Cloude 1987; Lüneburg 1995). Physically, a completely unpolarised wave has a polarisation which changes so radically in time that statistically any polarisation has the same amount of power as any other. In this situation, the wave can be represented by only one parameter (i.e. the amplitude of any component). The opposite case is when the $det([J]) = 0$ or $J_{HH}J_{VV} = J_{HV}J_{VH}$ and the wave is completely polarised. This is the stationary case, when the polarisation does not change in time and cross terms are exactly equal to the product of the two components. The latter expression could be seen as a Cauchy–Schwarz inequality which becomes equality for completely correlated components (Strang 1988). In general, cross terms increase with the polarisation purity of the wave.

The Stokes parameters are widely used to describe partial polarisations. They are initially defined in the case of pure polarisations:

$$\underline{q} = \begin{bmatrix} q_0 \\ q_1 \\ q_2 \\ q_3 \end{bmatrix} = \begin{bmatrix} |E_H|^2 + |E_V|^2 \\ |E_H|^2 - |E_V|^2 \\ |E_H||E_V| \cos \phi_{HV} \\ |E_H||E_V| \sin \phi_{HV} \end{bmatrix} = \begin{bmatrix} A^2 \\ A^2 \cos 2\chi \cos 2\psi \\ A^2 \cos 2\chi \sin 2\psi \\ A^2 \sin 2\chi \end{bmatrix}, \tag{3.13}$$

where A, ψ and χ are the ellipse parameters (Born and Wolf 1965; Goldstein and Collett 2003). As can be easily demonstrated the four Stokes parameters are not independent of each other, since $q_0^2 = q_1^2 + q_2^2 + q_3^2$. The parameters expressed in Eq. 3.13 are not sufficient to describe partial polarisations: they need to consider averaged components:

$$q = \begin{bmatrix} q_0 \\ q_1 \\ q_2 \\ q_3 \end{bmatrix} = \begin{bmatrix} \langle E_H E_H^* \rangle + \langle E_V E_V^* \rangle \\ \langle E_H E_H^* \rangle - \langle E_V E_V^* \rangle \\ \langle E_H E_V^* \rangle + \langle E_V E_H^* \rangle \\ j\langle E_H E_V^* \rangle - j\langle E_V E_H^* \rangle \end{bmatrix} = \begin{bmatrix} J_{HH} + J_{VV} \\ J_{HH} - J_{VV} \\ J_{HV} + J_{VH} \\ jJ_{HV} - iJ_{VH} \end{bmatrix}. \tag{3.14}$$

Equation 3.14 shows that the Stokes vectors can be easily associated with the Jones matrix elements (Zebker and Van Zyl 1991).

Two parameters are introduced to measure the polarimetric purity/impurity of the wave:

(a) The degree of coherence μ

$$\mu_{HV} = \frac{J_{HV}}{\sqrt{J_{HH} + J_{VV}}}, \tag{3.15}$$

which estimates the importance of the cross terms in the Jones matrix.

(b) The degree of polarisation D_ρ

$$D_\rho = \sqrt{1 - \frac{4Det([J])}{Trace([J])^2}} = \frac{\sqrt{q_1^2 + q_2^2 + q_3^2}}{q_0}, \tag{3.16}$$

which takes into account the correlation of the Jones vector components.

For a completely unpolarised wave $\mu_{HV} = D_\rho = 0$ and for a completely polarised one $\mu_{HV} = D_\rho = 1$.

The relationship $q_0^2 = q_1^2 + q_2^2 + q_3^2$ found for pure states of polarisation is not fulfilled for partial polarisations since $q_0^2 \geq q_1^2 + q_2^2 + q_3^2$. Physically, the variation of the polarisation ellipse (i.e. ψ and χ) reduces the last three elements of the Stokes vector but leaves unaffected the first element (related to the total power A^2).

The coherence matrix $[J]$ is based on the Jones vector and the unitary matrix (introduced previously) can be used with a similarity transformation to perform the change of basis (Van Zyl et al. 1987b):

$$[J_{ij}] = \left\langle ([U_2]\underline{E}_{ij})([U_2]\underline{E}_{ij})^{*T} \right\rangle = [U_2]\left\langle E_{ij}E_{ij}^{*T} \right\rangle [U_2]^{*T} = [U_2][J_{mn}][U_2]^{*T}. \tag{3.17}$$

3.2.4 Poincaré Polarisation Sphere

Considering that polarisation is basically a geometrical property of the plane wave, several techniques to visualise the field polarisation were introduced in the literature. The Poincaré sphere is one of the most powerful. This is based on a unique transformation from the space of the wave (two dimensional complex) to a three

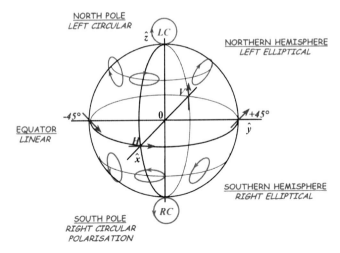

Fig. 3.2 Poincaré sphere (Boerner 2004) (Courtesy of Prof. E. Pottier and ESA)

dimensional real space (coordinate space) (Bebbington 1992; Lüneburg 1995; Ulaby and Elachi 1990).

Figure 3.2 represents the Poincaré sphere. The linear polarisations are on the equator, the left handed polarisations are on the upper hemisphere and the right handed on the lower one. The north and south poles represent respectively left and right circular polarisations. Any pure state of polarisation (i.e. $q_0^2 = q_1^2 + q_2^2 + q_3^2$) will be mapped on the surface of the sphere. Conversely, a partially polarised wave is inside the sphere (i.e. $q_0^2 \geq q_1^2 + q_2^2 + q_3^2$). A visual explanation is provided in (Deschamps and Edward 1973). The instant wave polarisation can be represented as a point that moves on the surface of the sphere. However, during the acquisition several polarisations states are averaged together leading to a resulting vector inside the sphere (i.e. the barycentre of points distributed on a sphere is inside the sphere).

3.2.5 Wave Decomposition Theorems

The wave coherence matrix is Hermitian, therefore the Single Value Decomposition (SVD) can be applied and it has real positive eigenvalues (Strang 1988). In particular, the diagonalisation extracts a basis (the eigenvectors) where the representation of the matrix is diagonal.

$$\begin{bmatrix} J_{mm} & J_{mn} \\ J_{nm} & J_{nn} \end{bmatrix} = \begin{bmatrix} e_{11} & e_{12} \\ e_{21} & e_{22} \end{bmatrix} \begin{bmatrix} \lambda_1 & 0 \\ 0 & \lambda_2 \end{bmatrix} \begin{bmatrix} e_{11} & e_{12} \\ e_{21} & e_{22} \end{bmatrix}^{*T}. \tag{3.18}$$

The eigenvalues are $\lambda_1 \geq \lambda_2 \geq 0$, and the eigenvector $\underline{e}_1 = [e_{11}, e_{12}]^T$, $\underline{e}_2 = [e_{22}, e_{21}]^T$ are unitary vectors forming the columns of a unitary (full rank) matrix that performs a rotation of $[J]$. The diagonalisation is a unique procedure and the eigenvalues are invariant of $[J]$. In particular, the SVD has remarkable physical insight since the two eigenvectors represent two orthogonal axes where the wave is completely polarised (Goldstein and Collett 2003; Cloude 2009; Lee and Pottier 2009).

$$[J_1] = \left(\underline{e}_1 \underline{e}_1^{*T}\right) \text{ and } [J_2] = \left(\underline{e}_2 \underline{e}_2^{*T}\right). \tag{3.19}$$

Two equivalent decompositions can be performed:

(a) Two contributions, one completely polarised and one completely unpolarised:

$$[J] = (\lambda_1 - \lambda_2)[J_1] + \lambda_2[I_2]. \tag{3.20}$$

The second component can be interpreted as polarimetric noise, since its two eigenvectors are equal. Please note, thermal noise is only an example of polarimetric noise and several scatterers have this behaviour.

(b) Two orthogonal, completely polarised contributions:

$$[J] = \lambda_1[J_1] + \lambda_2[J_2]. \tag{3.21}$$

The two contributions represent pure states of wave polarisation since the matrices $[J_1]$ and $[J_2]$ have rank one (i.e. the columns are dependent). Their determinants will vanish $Det([J_1]) = Det([J_2]) = 0$, leading to the relation found earlier for completely polarised waves.

In conclusion, a plane wave can always be represented by the composition of two completely polarised waves. In the case $\lambda_1 = \lambda_2$, the two waves statistically have the same power and the total wave is completely unpolarised. On the other hand, if $\lambda_2 = 0$ we can express the total wave with just one polarised state (i.e. the wave is completely polarised). Consequently, the eigenvalues can be used to extract information regarding the degree of polarisation of the wave:

$$D_\rho = \frac{\lambda_1 - \lambda_2}{\lambda_1 + \lambda_2}. \tag{3.22}$$

As expected, the degree of polarisation (which is a physical property of the wave) can be expressed with invariants of the coherence matrix (i.e. the eigenvalues), hence it is an invariant itself (Cloude 2009).

Another parameter widely utilised to extract information about the polarimetric purity is the wave entropy:

$$H = \sum_{i=1}^{2} (-P_i \log_2 P_i), \quad \text{with} \quad P_i = \frac{\lambda_i}{\lambda_1 + \lambda_2}, \tag{3.23}$$

where P_i represents a probability, hence, $P_1 + P_2 = 1$. The wave entropy H varies between 0 and 1 and provides information about the randomness of the polarisation state. Specifically, when $H = 1$ the two eigenvalues are equal and the wave is completely unpolarised. On the other hand, when $H = 0$ the second eigenvalue is zero and the wave is completely polarised.

3.3 Target Polarimetry: Single Targets

Wave polarimetry builds up the fundaments to describe the polarimetric behaviour of objects illuminated by an EM wave. However, one single wave scattered from a target is insufficient to characterise completely and uniquely the target. As explained in the previous chapter, several measurements must be performed. The starting point of this section is the concept of the scattering matrix.

Similarly to the case of wave polarimetry, we structured the treatment making a separation between targets which can be represented completely by a single scattering matrix (therefore called *single* targets) and the rest (i.e. *partial* targets) (Ulaby and Elachi 1990). In principle, the use of a unique scattering matrix is feasible if and only if the target polarimetric behaviour does not change in time/space (i.e. polarimetric stationarity). For instance, a steady object illuminated by a completely polarised wave scatters a wave which is completely polarised (i.e. $D_\rho = 1$). Moreover, different realisations of the same target have to scatter the same polarisation. For some typologies of single targets the hypothesis of polarised illumination can be relaxed, since the scattered wave is always completely polarised (therefore they are regarded as *polarisers*) (Born and Wolf 1965; Goldstein and Collett 2003).

The counterparts of single targets are the partial targets. During the acquisition the time is fixed (the samples are acquired in a precise timestamp), therefore the variation in polarisation states is provided by the spatial difference. A partial target cannot be characterised by a single pixel acquisition since for a stochastic process any realisation can be different from the others. In order to extract useful information an ensemble average must be performed (Oliver and Quegan 1998). Commonly, partial targets are classified as distributed targets, which are composed of several scatterers. The equivalence of these two target typologies is frequently verified in real data. However, some exceptional distributed targets can be described by a single scattering matrix. For instance, the latter can be composed by a collection of equal polarisers (e.g. a Yagi antenna) (Collin 1985). We will come back to this concept during the validation of the detector (Chap. 6).

3.3.1 *Sinclair Matrix and Basis Transformation*

The derivation of the scattering (Sinclair) matrix is illustrated in the first chapter. The scattering (or Sinclair) matrix expressed in linear horizontal and vertical basis is

$$
\begin{bmatrix} E_H^s \\ E_V^s \end{bmatrix} = \frac{1}{\sqrt{4\pi r^2}} \begin{bmatrix} S_{HH} & S_{HV} \\ S_{VH} & S_{VV} \end{bmatrix} \begin{bmatrix} E_H^i \\ E_V^i \end{bmatrix}.
\tag{3.24}
$$

where E_H^s, E_V^s are the scattered and E_H^i, E_V^i are the incident waves (Kennaugh and Sloan 1952; Kostinski and Boerner 1986; Krogager 1993; Sinclair 1950). Algebraically, the scattering matrix can be interpreted as a transformation from an incident to a scattered wave and it is necessary and sufficient to characterise the polarimetric behaviour of a single target (Cloude 1995c, 1986). In particular, under the hypotheses of (i) monostatic sensor (same transmitter and receiver antenna) and (ii) medium reciprocity, the two off-diagonal terms are identical $S_{HV} = S_{VH}$ with the exception of the noise. This property can be seen as an extension of the reciprocal theorem for antennas (an antenna has the same behaviour in transmission and reception) (Collin 1985).

The scattering matrix is dependent on the basis (i.e. coordinates) chosen for the acquisition: in our case the linear horizontal and vertical polarisations. However, the physical insight cannot vary when rotating the axis used to acquire the measurements. As for the wave polarisation counterpart an operation must be introduced to perform the change of basis of the scattering matrix:

$$
[S]_{ij} = [U_2][S]_{HV}[U_2]^T.
\tag{3.25}
$$

where again $[U_2]$ is a unitary matrix (Lüneburg 1995). The link between the change of basis and the complex polarisation ratio ρ in the new basis is

$$
S_{ii} = \frac{1}{1 + \rho\rho^*} \left[S_{HH} - \rho^* S_{HV} - \rho^* S_{VH} + \rho^{*2} S_{VV} \right],
$$

$$
S_{ij} = \frac{1}{1 + \rho\rho^*} \left[\rho S_{HH} + S_{HV} - \rho\rho^* S_{VH} - \rho^* S_{VV} \right],
$$

$$
S_{ji} = \frac{1}{1 + \rho\rho^*} \left[\rho S_{HH} - \rho\rho^* S_{HV} + S_{VH} - \rho^* S_{VV} \right],
\tag{3.26}
$$

$$
S_{jj} = \frac{1}{1 + \rho\rho^*} \left[\rho^2 S_{HH} + \rho S_{HV} - \rho S_{VH} + S_{VV} \right],
$$

$$
\rho = \frac{\cos 2\chi \sin 2\psi + j \sin 2\chi}{1 + \cos 2\chi \cos 2\psi}.
$$

In order to extract physical information from $[S]$, we are interested in the invariants of the matrix. For instance its determinant

$$Det([S]_{HV}) = Det([S]_{ij}). \qquad (3.27)$$

Please note, the unitary matrix used in the change of basis does not change the determinant.

Additionally, the total power acquired in the polarimetric measurement is invariant as well. It can be calculated with the span of the scattering matrix.

$$Span([S]) = |S_{HH}|^2 + |S_{HV}|^2 + |S_{VH}|^2 + |S_{VV}|^2 = |S_{ii}|^2 + |S_{ij}|^2 + |S_{ji}|^2 + |S_{jj}|^2. \quad (3.28)$$

3.3.2 Scattering Features Vectors

The aim of this section is to provide a geometrical representation of the target based on a vector rather than a matrix. The reason is that with vectors it is often easier to handle algebraic manipulations (Cloude 1987; Ulaby and Elachi 1990). A scattering features vector was introduced:

$$\underline{k}_4 = \frac{1}{2} Trace\{[S]\Psi\} = [k_1, k_2, k_3, k_4]^T. \qquad (3.29)$$

where Ψ is a complete set of 2×2 complex basis matrices under a Hermitian inner product. Considering Ψ is a complete basis set for the matrix space, all the information kept in the scattering matrix are reversed into the scattering vector. Algebraically, the procedure can be interpreted as a rearrangement of the polarimetric information through linear combinations of the scattering matrix elements. Therefore, the two representations are completely equivalent (Strang 1988).

In the literature, two standard basis sets have been utilised (Boerner 2004; Touzi et al. 2004):

(a) Lexicographic basis:

$$[\Psi_L] = \left\{ 2\begin{bmatrix} 1 & 0 \\ 0 & 0 \end{bmatrix}, 2\begin{bmatrix} 0 & 1 \\ 0 & 0 \end{bmatrix}, 2\begin{bmatrix} 0 & 0 \\ 1 & 0 \end{bmatrix}, 2\begin{bmatrix} 0 & 0 \\ 0 & 1 \end{bmatrix} \right\}, \qquad (3.30)$$

where the resulting feature vector is

$$\underline{k}_{4L} = [S_{HH}, S_{HV}, S_{VH}, S_{VV}]^T. \qquad (3.31)$$

This representation is advantageous since in some situations it can simplify the calculations. Besides, the elements are related to special targets (respectively horizontal dipoles, 45° oriented dihedral and vertical dipoles).

(b) Pauli basis:

The spin basis introduced by Pauli and adapted to the BSA coordinate systems (please note the Pauli basis has a different expression in the FSA system) are (Cloude 1987):

$$[\Psi_P] = \left\{ \sqrt{2} \begin{bmatrix} 1 & 0 \\ 0 & 1 \end{bmatrix}, \sqrt{2} \begin{bmatrix} 1 & 0 \\ 0 & -1 \end{bmatrix}, \sqrt{2} \begin{bmatrix} 0 & 1 \\ 1 & 0 \end{bmatrix}, \sqrt{2} \begin{bmatrix} 0 & -j \\ j & 0 \end{bmatrix} \right\}, \quad (3.32)$$

and the Pauli scattering vector is

$$\underline{k}_{4P} = [S_{HH} + S_{VV}, S_{HH} - S_{VV}, S_{HV} + S_{VH}, j(S_{HV} - S_{VH})]^T. \quad (3.33)$$

The benefit of the Pauli representation is the direct association with physical targets. In particular the first element $S_{HH} + S_{VV}$ represents isotropic scatterers like spheres and surfaces (also regarded as odd bounce), $S_{HH} - S_{VV}$ is related to dihedral with horizontal corner line between the two plates (also named even bounce), $S_{HV} + S_{VH}$ is a dihedral with the corner 45° oriented and $j(S_{HV} - S_{VH})$ is a non reciprocal target (Cloude 2009; Lee and Pottier 2009; Ulaby and Elachi 1990; Zebker and Van Zyl 1991). Based on the simple physical interpretation, the Pauli scattering vector can be used as a coherent decomposition of the observed target.

3.3.3 Backscattering Case

As mentioned previously, in the case of (i) monostatic sensor and (ii) reciprocal medium, the scattering matrix is symmetric. In physics symmetries are often related to a significant simplification of the scattering problem, consequently we want to adopt them in our treatment (Cloude 1995b, c). In fact, SAR polarimetric acquisitions commonly exploit monostatic sensors (backscattering problem) and at microwave radiation observed targets are generally reciprocal. An exception is the satellite observations in low frequency, where the ionosphere can be non reciprocal due to the presence of plasma (i.e. Faraday rotation) (Freeman 1992).

When case (i) and (ii) are fulfilled, the two off-diagonal terms of the scattering matrix are equal, $S_{HV} = S_{VH}$ (with the exception of noise). As a consequence, 3 rather than 4 complex numbers are necessary to characterise the target. The Pauli and lexicographic scattering vectors can be rewritten as

$$\underline{k}_L = \left[S_{HH}, \sqrt{2}S_{HV}, S_{VV} \right]^T,$$

$$\underline{k}_P = \frac{1}{\sqrt{2}} [S_{HH} + S_{VV}, S_{HH} - S_{VV}, 2S_{HV}]^T. \quad (3.34)$$

where the factors introduced are necessary to keep the span invariant. Please, note the non-reciprocal component in the Pauli scattering vector is removed.

The scattering vector depends on the basis set used in the feature vector creation. Additionally, $[S]$ itself depends on the basis exploited to acquire the polarimetric data.

The relationship between the Pauli and Lexicographic scattering vector is (Boerner et al. 1997):

$$\underline{k}_P = [D_3]\underline{k}_L,$$

$$\underline{k}_L = [D_3]^{-1}\underline{k}_P. \tag{3.35}$$

$[D_3]$ is a transformation where the new basis are represented by its columns:

$$[D_3] = \frac{1}{\sqrt{2}}\begin{bmatrix} 1 & 0 & 1 \\ 1 & 0 & -1 \\ 0 & \sqrt{2} & 0 \end{bmatrix}. \tag{3.36}$$

The factor $\dfrac{1}{\sqrt{2}}$ keeps the span invariant.

Regarding the basis of $[S]$, the operation is accomplished by multiplying by a unitary matrix:

$$\underline{k}_L(AB) = [U_{3L}(\rho)]\underline{k}_L(HV),$$

$$\underline{k}_P(AB) = [U_{3P}(\rho)]\underline{k}_P(HV). \tag{3.37}$$

The complex polarisation ratio ρ can be used to define the terms of the unitary rotation

$$[U_{3L}] = \frac{1}{1+\rho\rho^*}\begin{bmatrix} 1 & \sqrt{2}\rho & \rho^2 \\ -\sqrt{2}\rho^* & 1-\rho\rho^* & \sqrt{2}\rho \\ \rho^{*2} & -\sqrt{2}\rho^* & 1 \end{bmatrix},$$

$$[U_{3P}] = \frac{1}{2(1+\rho\rho^*)}\begin{bmatrix} 2+\rho^2+\rho^{*2} & \rho^{*2}-\rho^2 & 2(\rho-\rho^*) \\ \rho^2-\rho^{*2} & 2-(\rho^2+\rho^{*2}) & 2(\rho+\rho^*) \\ 2(\rho-\rho^*) & -2(\rho+\rho^*) & 2(1-\rho\rho^*) \end{bmatrix}, \tag{3.38}$$

where $Det([U_{3L}]) = Det([U_{3P}]) = 1$.

The change of basis must preserve the span of $[S]$, which is equivalent to the norm of the scattering vector:

$$\|\underline{k}\| = \frac{1}{2}Span([S]) = \frac{1}{2}Trace([S][S]^*) = |S_{HH}|^2+|S_{HV}|^2+|S_{VH}|^2+|S_{VV}|^2. \tag{3.39}$$

Starting from the scattering feature vector it is possible to derive the scattering mechanism which is a normalised vector keeping the polarimatric information:

$$\underline{\omega} = \frac{\underline{k}}{\|\underline{k}\|}. \tag{3.40}$$

The unitary scattering mechanism can be exploited to extract the projection of the polarimetric data over a particular target of interest:

$$i(\underline{\omega}) = \underline{\omega}^{*T}\underline{k}. \tag{3.41}$$

The projection is a complex scalar and can be interpreted as a SAR image.

Now it is possible to define the normalised cross correlation between projections over two different scattering mechanisms (Cloude 2009). This is named polarimetric coherence:

$$\gamma = \frac{\langle i(\underline{\omega}_1)\, i(\underline{\omega}_2)^* \rangle}{\sqrt{\langle i(\underline{\omega}_1)\, i(\underline{\omega}_1)^* \rangle \langle i(\underline{\omega}_2)\, i(\underline{\omega}_2)^* \rangle}}. \tag{3.42}$$

3.3.4 Polarisation Fork

One fascinating topic in radar polarimetry is the assessment of the polarisations providing the maximum return from a given single target (Huynen 1970; Kennaugh and Sloan 1952; Kennaugh 1981). Once the optimum polarisation is known, the antenna can be tuned to it in order to improve the detection performance. Besides, it is relevant to find the polarisation able to delete completely the return coming from a clutter source (for instance clouds in aircraft detection). A series of experiments and theoretical work led to the formulation of the Polarisation Nulls theory (Agrawal and Boerner 1989; Boerner et al. 1981; Boerner 1981; Huynen 1970; Kennaugh and Sloan 1952; Cloude 1987).

In particular, four characteristic polarisations were initially identified.

(a) 2 optimum polarisations or Cross-pol Nulls:

If a target is illuminated by a Cross-pol Null, the backscattering in the cross polarisation vanishes. In particular, the entire energy scattered backward is concentrated in the co-polarisations. The cross terms of the scattering matrix vanish and this becomes diagonal. In other words, the Cross-pol Nulls are the eigenvectors which diagonalise $[S]$ while their backscattering are the eigenvalues. Please note, the scattering matrix is symmetric, hence it can be diagonalised and the eigenvalues will generally be complex numbers. The first eigenvector (the one with highest eigenvalue) represents the polarisation with the maximum return from the target.

Alternative ways to calculate the Cross-pol Nulls consider the diagonalisation of the Graves matrix (Graves 1956) (which will be introduced in the next section) or performing an optimisation of the co-polarisations with the Lagrangian approach. These polarisations will be regarded in the following as X_1 and X_2.

Fig. 3.3 Polarisation fork

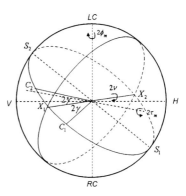

(b) 2 Co-pol Nulls:

These are polarisations that when transmitted do not have any return in the co-polarisation since all the backscattered energy is located in the cross polarisation. If a Co-pol Nulls basis is exploited to acquire the scattering matrix, the diagonal terms will vanish. There exist several ways to calculate the Co-pol Nulls. The use of the Lagrangian over all the possible cross polarisations is the most used approach (Boerner et al. 1981).

The practical relevance of the Co-pol Nulls is that they can be employed to mask out clutter. Unfortunately, the Co-pol Nulls lose significance for partial targets since the exact null of the backscattering is never obtained. In the following, these polarisation are regarded as C_1 and C_2.

If Cross-pol Nulls and Co-pol Nulls are displayed on the Poincaré sphere they will lie on the same plane shaping a fork as shown in Fig. 3.3. The Cross-pol Nulls X_1 and X_2 are antipodal points on the sphere, since they are orthogonal polarisations (Huynen 1970; Kennaugh and Sloan 1952). Regarding the Co-pol Nulls C_1 and C_2, they have the same angular distance from the maximum Cross-pol Null, 2γ (the physical meaning of the angle γ will be explained in the following). Unfortunately, in the case of partial targets, the four characteristic polarisations do not lie on the sphere (or even the same plane) and the whole concept of the polarisation fork loses relevance since it cannot be rigorously defined (Boerner et al. 1991; Van Zyl et al. 1987a).

Any single target has a unique polarisation fork. In the following, two special cases largely used in the validation of the proposed detector are illustrated (Cloude 1987; Huynen 1970):

(a) Reflections: the two eigenvalues of the scattering matrix have the same amplitude. In this instance, the two Co-pol Nulls C_1 and C_2 become antipodal on the Poincaré sphere (this can be proved with the Huynen parameters presented in the next section). Cross-pol null couples are infinite in number and they lie on the circle made by the interception of the Poincarè sphere with the plane passing through the centre and the polarisation fork.

(b) Degenerate: if there is only one eigenvalue different from zero, the eigen-problem is degenerate (the rank of the scattering matrix is 1). In this situation the two Co-pol Nulls C_1 and C_2 will be coincident and antipodal to the first Cross-pol Null X_1. Hence C_1 and C_2 and X_2 will lie on the same point on the Poincaré sphere. Examples of degenerate targets are dipoles.

After the pioneering work of Kennaugh and Huynen (who first studied the characteristic polarisations), other work was carried out on the polarisations able to characterise the polarimetric behaviour of targets (Boerner et al. 1981; Boerner 1981; Xi and Boerner 1992). Four other polarisations were added to the previous list:

(c) 2 Cross-pol Maxima: these are polarisations that when transmitted have the maximum return for the cross polarisation. They are obtained with a maxi-misation of the cross terms (for instance with a Lagrangian approach). These polarisations lie on the plane of the polarisation fork but generally they are different from the Co-pol Nulls. They are antipodal to each other and have an angular distance of 90° from the Cross-pol Nulls. They are indicated with S_1 and S_2 on Fig. 3.3. Cross-pol Maxima overlap with Co-pol Nulls in the case of multiple reflections (i.e. when the two eigenvalues of the scattering matrix are the same), since the angle between Co-pol Nulls and Cross-pol Nulls in this case is 90°.

(d) 2 Cross-pol Saddle points: these points have no strong physical meaning, but geometrical meaning on the Poincarè sphere, since they lie on the interception with the normal of the polarisation fork passing through the centre. (Please note, for the sake of readability in Fig. 3.3 the Saddle points are not depicted).

3.3.5 Huynen Single Target Decomposition

The polarisation fork is particularly advantageous for radar systems designed to focus on a particular target, since it is a collection of actual polarisations. On the other hand, in this formalism the link with the physics of the scattering could be lost. A parameterisation can be helpful to improve this link. In this section, the Huynen parameterisation of the scattering matrix or Huynen coherent decomposition (not to be confused with the Huynen incoherent decomposition) is presented (Huynen 1970; Pottier 1992). As introduced previously, 6 real parameters are needed to completely describe the scattering matrix. Huynen developed a representation based on invariants of the scattering matrix, as the eigenvectors (i.e. Cross-pol Nulls).

The expression of the scattering matrix with the Huynen parameters is

$$[S] = [R(\psi_m)][T(\chi_m)][S_d][T(\chi_m)][R(-\psi_m)],$$

$$[S_d] = \begin{pmatrix} me^{i(v+\zeta)} & 0 \\ 0 & m\tan(\gamma)e^{-i(v-\zeta)} \end{pmatrix},$$

$$[T(\tau_m)] = \begin{pmatrix} \cos\chi_m & -i\sin\chi_m \\ -i\sin\chi_m & \cos\chi_m \end{pmatrix}, \qquad (3.43)$$

$$[R(\phi_m)] = \begin{pmatrix} \cos\psi_m & -\sin\psi_m \\ \sin\psi_m & \cos\psi_m \end{pmatrix}.$$

The maximum eigenvector can be represented by four real numbers: the amplitude m, the absolute phase ζ, the orientation ψ_m and the ellipticity angle χ_m. Once the first eigenvector is fixed, the second one is obtained as the antipodal polarisation on the Poincarè sphere. ψ_m has a central role in the representation since it makes the rest of the parameters independent of the orientation angle (rotations around the LOS: Line of Sight). The last two parameters are related to the amplitude and phase relationship between the two eigenvalues. Specifically, γ is the characteristic angle defining the reciprocal weight of the eigenvectors. It controls the target typology, moving from multiple reflections for $\gamma = 45°$ to degenerate target (e.g. dipole) for $\gamma = 0°$. Finally, the skip angle v is associated with the eigenvalues phase relationship. It is named skip angle because in the case of multiple reflections it determines if the number of reflections is even or odd (Huynen 1970).

The absolute phase ζ needs a further clarification. This angle contains information about a physical property of the target, however in single pass polarimetry it cannot be separated from the phase due to the range distance (i.e. geometrical phase). In other words, if the target moves along the range the phase changes even though the target remains the same. Therefore, ζ is generally neglected in single pass polarimetry.

Regarding the amplitude m, in the case of scattering mechanism this is always equal to 1 since the vector is normalised. In conclusion, the scattering mechanism has only 4° of freedom represented by ψ_m, χ_m, γ and v.

3.4 Target Polarimetry: Partial Targets

A partial target scatters an EM wave with a degree of polarisation smaller than 1. Considering that the polarisation state is a function of time/space, partial targets can be modelled as random processes. Dealing with random processes, a single realisation is not sufficient to describe the process completely, since distinct realisations can be drastically different. A statistical description is required, specifically the second order moments of the scattering vector components can be estimated (Dong and Forster 1996; Lang 1981; Oliver and Quegan 1998).

3.4.1 *Muller and Graves Matrices*

In the previous section, the Stokes parameters were used to describe a partial state of wave polarisation. However, they are not sufficient for partial targets since they are three independent real numbers, while the target has more degrees of freedom (e.g. a single target has 5). In other words, more than one Stokes vector is required. The scattering process can be interpreted as a transformation between an incident and a scattered wave. In the previous section, these two waves were polarised and to a 2×2 scattering matrix. On the other hand, partial targets scatter partially polarised waves, hence the Stokes vector must be utilised (Beckmann 1968).

In conclusion, partial target scattering can be explained as a transformation between an incident and a scattered Stokes vector. This transformation can be represented with a 4×4 real matrix which is regarded as the Muller matrix $[M]$ (Barakat 1981; Cloude 1986; Kennaugh and Sloan 1952; Van der Mee and Hovenier 1992). $[M]$ is 4×4 with real entries, therefore has 16 real numbers (they are all independent only in bistatic case). In the literature, the Muller matrix applied on monostatic and reciprocal medium is sometimes called the Kennaugh matrix $[K]$.

The transformation can be written as:

$$\underline{q}^s = [M]\underline{q}^i. \tag{3.44}$$

Single targets represent a special case where only 6 of the 16 terms are independent and there exists a unique link between $[M]$ and $[S]$.

The Muller matrix is not the only technique exploiting the second order statistics of the partial target. In the literature, the power matrix or Graves matrix (Graves 1956) was proposed:

$$[G] = \langle [S]^{*T}[S] \rangle = \begin{bmatrix} K_{11} + K_{12} & K_{13} - jK_{14} \\ K_{13} + jK_{14} & K_{11} - K_{12} \end{bmatrix}. \tag{3.45}$$

$[G]$ is positive semi-definite and diagonalisable. Its eigenvectors are the optimum polarisations for the scattering matrix (since the eigenvectors do not change when a matrix is squared).

The Graves matrix leads to an incoherent target decomposition able to separate the power terms due to horizontal and vertical components:

$$[G] = [G_H] + [G_V] = \left\langle \begin{bmatrix} |S_{HH}|^2 & S_{HV}S_{HH}^* \\ S_{HH}S_{HV}^* & |S_{HV}|^2 \end{bmatrix} \right\rangle + \left\langle \begin{bmatrix} |S_{VH}|^2 & S_{VV}S_{VH}^* \\ S_{VH}S_{VV}^* & |S_{VV}|^2 \end{bmatrix} \right\rangle.$$
$$\tag{3.46}$$

3.4.2 Covariance Matrix

Another common way to estimate the second order statistics exploits the scattering vector (Cloude 1987; Lee and Pottier 2009; Ulaby and Elachi 1990; Zebker and Van Zyl 1991):

$$[C_{4L}] = \langle \underline{k}_{4L}\ \underline{k}_{4L}^{*T} \rangle = \begin{bmatrix} \langle |S_{HH}|^2 \rangle & \langle S_{HH}S_{HV}^* \rangle & \langle S_{HH}S_{VH}^* \rangle & \langle S_{HH}S_{VV}^* \rangle \\ \langle S_{HV}S_{HH}^* \rangle & \langle |S_{HV}|^2 \rangle & \langle S_{HV}S_{VH}^* \rangle & \langle S_{HV}S_{VV}^* \rangle \\ \langle S_{VH}S_{HH}^* \rangle & \langle S_{VH}S_{HV}^* \rangle & \langle |S_{VH}|^2 \rangle & \langle S_{VH}S_{VV}^* \rangle \\ \langle S_{VV}S_{HH}^* \rangle & \langle S_{VV}S_{HV}^* \rangle & \langle S_{VV}S_{VH}^* \rangle & \langle |S_{VV}|^2 \rangle \end{bmatrix}. \quad (3.47)$$

The resulting matrix is a standard covariance matrix, where the random variables have zero mean. In the case of a monostatic sensor and reciprocity it becomes:

$$[C_L] = \langle \underline{k}_L\ \underline{k}_L^{*T} \rangle = \begin{bmatrix} \langle |S_{HH}|^2 \rangle & \sqrt{2}\langle S_{HH}S_{HV}^* \rangle & \langle S_{HH}S_{VV}^* \rangle \\ \sqrt{2}\langle S_{HV}S_{HH}^* \rangle & 2\langle |S_{HV}|^2 \rangle & \sqrt{2}\langle S_{HV}S_{VV}^* \rangle \\ \langle S_{VV}S_{HH}^* \rangle & \sqrt{2}\langle S_{VV}S_{HV}^* \rangle & \langle |S_{VV}|^2 \rangle \end{bmatrix}. \quad (3.48)$$

Equivalently, the covariance matrix can be estimated starting from the Pauli scattering vector. For brevity only the 3×3 scenario is presented. This is regarded as coherency matrix $[T]$:

$$[C_P] = [T] = \langle \underline{k}_P\ \underline{k}_P^{*T} \rangle$$
$$= \begin{bmatrix} \langle |S_{HH}|^2 + |S_{HH}|^2 \rangle & \langle (S_{HH}+S_{VV})(S_{HH}-S_{VV})^* \rangle & 2\langle (S_{HH}+S_{VV})(S_{HV})^* \rangle \\ \langle (S_{HH}+S_{VV})^*(S_{HH}-S_{VV}) \rangle & \langle |S_{HH}|^2 - |S_{HH}|^2 \rangle & 2\langle (S_{HH}-S_{VV})(S_{HV})^* \rangle \\ 2\langle (S_{HH}+S_{VV})^*(S_{HV}) \rangle & 2\langle (S_{HH}-S_{VV})^*(S_{HV}) \rangle & 4\langle |S_{HV}|^2 \rangle \end{bmatrix}. \quad (3.49)$$

By definition, the covariance matrix $[C]$ can be estimated starting from any basis set. If the scattering vector in a generic basis is given by $\underline{k} = [k_1, k_2, k_3]^T$ then:

$$[C] = \langle \underline{k}\ \underline{k}^{*T} \rangle = \begin{bmatrix} \langle |k_1|^2 \rangle & \langle k_1 k_2^* \rangle & \langle k_1 k_3^* \rangle \\ \langle k_2 k_1^* \rangle & \langle |k_2|^2 \rangle & \langle k_2 k_3^* \rangle \\ \langle k_3 k_1^* \rangle & \langle k_3 k_2^* \rangle & \langle |k_3|^2 \rangle \end{bmatrix}. \quad (3.50)$$

In the following the symbol $[C]$ will be used to describe a covariance matrix independently of the selected basis.

The elements on the diagonal are real positive (regarded as powers). The sum of the diagonal elements (i.e. *Trace* of the matrix) is a polarimetric invariant since it represents the total power acquired by the system or the span of the scattering matrix:

$$Trace([C]) = Span([S]). \tag{3.51}$$

The diagonal terms are the cross-correlations between the components of the scattering vector. They provide information about the presence of coherent targets or in general the degree of polarisation.

As for scattering vectors, the basis can be modified with unitary transformations:

$$\begin{aligned} [C(AB)] &= \left\langle [U_{3L}]\underline{k}_L(HV)([U_{3L}]\underline{k}_L(HV))^{*T} \right\rangle \\ &= [U_{3L}]\left\langle \underline{k}_L(HV)\underline{k}_L(HV)^{*T} \right\rangle [U_{3L}]^{*T} = [U_{3L}][C(HV)][U_{3L}]^{*T}. \end{aligned} \tag{3.52}$$

Obviously, the change of basis does not modify the total power backscattered:

$$Trace([C(AB)]) = Trace([C(HV)]). \tag{3.53}$$

Considering the Hermitian symmetry, only 3 real (i.e. the diagonal) and 3 complex (i.e. off-diagonal of the upper triangular part) terms are independent of each other. Therefore, 9 real parameters are necessary and sufficient to characterise partial targets (Cloude 1995c, 1986).

Starting from the covariance matrix, it is possible to define the polarimetric coherence between two scattering mechanisms $\underline{\omega}_1$ and $\underline{\omega}_2$ as:

$$\begin{aligned} \gamma &= \frac{\left\langle i(\underline{\omega}_1)\, i(\underline{\omega}_2)^* \right\rangle}{\sqrt{\left\langle i(\underline{\omega}_1)\, i(\underline{\omega}_1)^* \right\rangle \left\langle i(\underline{\omega}_2)\, i(\underline{\omega}_2)^* \right\rangle}} \\ &= \frac{\left\langle \left(\underline{\omega}_1^{*T}\underline{k} \right) \left(\underline{k}^{*T}\underline{\omega}_2 \right)^* \right\rangle}{\sqrt{\left\langle \left(\underline{\omega}_1^{*T}\underline{k} \right) \left(\underline{k}^{*T}\underline{\omega}_1 \right)^* \right\rangle \left\langle \left(\underline{\omega}_2^{*T}\underline{k} \right) \left(\underline{k}^{*T}\underline{\omega}_2 \right)^* \right\rangle}} \\ &= \frac{\underline{\omega}_1^{*T}\left\langle \underline{k}\,\underline{k}^{*T} \right\rangle \underline{\omega}_2}{\sqrt{\left(\underline{\omega}_1^{*T}\left\langle \underline{k}\,\underline{k}^{*T} \right\rangle \underline{\omega}_1 \right) \left(\underline{\omega}_2^{*T}\left\langle \underline{k}\,\underline{k}^{*T} \right\rangle \underline{\omega}_2 \right)}}. \end{aligned} \tag{3.54}$$

Hence,

$$\gamma(\underline{\omega}_1, \underline{\omega}_2) = \frac{\left| \underline{\omega}_1^{*T}[C]\underline{\omega}_2 \right|}{\sqrt{\left(\underline{\omega}_1^{*T}[C]\underline{\omega}_1 \right) \left(\underline{\omega}_2^{*T}[C]\underline{\omega}_2 \right)}}. \tag{3.55}$$

3.4.3 Eigenvalue Decomposition (Cloude–Pottier)

By definition, the covariance matrix is semi-definite positive and Hermitian. Therefore, it can be diagonalised and the eigenvalues are real positive (Cloude 1992; Cloude and Pottier 1996; Van Zyl 1992).

The eigenproblem to solve is (Strang 1988):

$$[T]\underline{u}_i = \lambda_i \underline{u}_i \quad \text{for} \quad i = 1, 2, 3 \tag{3.56}$$

or equivalently

$$([T] - \lambda_i[I])\underline{u}_i = \underline{0} \quad \text{for} \quad i = 1, 2, 3 \tag{3.57}$$

The three resulting eigenvectors $\underline{u}_1$, $\underline{u}_2$ and $\underline{u}_3$ represent a basis where the components are independent of each other. The change of basis which makes $[T]$ diagonal can be accomplished with a unitary matrix $[U] = [\underline{u}_1, \underline{u}_2, \underline{u}_3]$, with the eigenvectors as columns:

$$\begin{aligned} [T] &= \left\langle [U]\underline{k}_{eigen} ([U]\underline{k}_{eigen})^{*T} \right\rangle \\ &= [U]\left\langle \underline{k}_{eigen}\, \underline{k}_{eigen}^{*T} \right\rangle [U]^{*T} = [U][\Sigma][U]^{*T}. \end{aligned} \tag{3.58}$$

where $[\Sigma]$ is the diagonal matrix of the eigenvalues λ_1, λ_2 and λ_3:

$$[\Sigma] = \begin{bmatrix} \lambda_1 & 0 & 0 \\ 0 & \lambda_2 & 0 \\ 0 & 0 & \lambda_3 \end{bmatrix} = diag(\lambda_1, \lambda_2, \lambda_3). \tag{3.59}$$

Once the eigenvector basis is extracted, $[T]$ can be decomposed into three independent contributions:

$$[T] = \sum_{i=1}^{3} \lambda_i [T_i] = \sum_{i=1}^{3} \lambda_i \underline{u}_i\, \underline{u}_i^{*T}. \tag{3.60}$$

Each contribution is a target with a coherence matrix of rank one, therefore a single target. Remarkably, any coherence matrix is Hermitian and the eigenvalues are independent of the basis. As a result, the decomposition can be applied to any partial target and it is unique (Cloude 1992, 1995b).

The diagonalised matrix $[\Sigma]$ is obtained with a unitary transformation (specifically a similarity), consequently, the sum of the eigenvalues is equal to the span of the scattering matrix:

$$\lambda_1 + \lambda_2 + \lambda_3 = Trace([\Sigma]) = Span([S]). \tag{3.61}$$

The eigenvalues are associated with the power scattered by a single target. Once sorted, λ_1 represents the strongest single target. On the other hand, λ_3 is associated with the single target with minimum return (in general this is just an algebraic rather than a real target in the scene). When $\lambda_1 \neq 0$ and $\lambda_2 = \lambda_3$ the $[T]$ matrix itself has rank one and only one single target is present in the scene. Clearly, the latter occurrence is just theoretical since the thermal noise is spread over all the components and $[T]$ will always be full rank. On the other hand, when $\lambda_1 = \lambda_2 = \lambda_3$ any single targets in the scene share the same amount of backscattering (this is the counterpart of a completely unpolarised wave in target polarimetry).

It is apparent that the reciprocal weight among the eigenvalues is related to the target degree of polarisation. A methodology analogous to the wave entropy can be exploited extracting the target entropy H:

$$H = \sum_{i=1}^{3} (-P_i \log_3 P_i), \tag{3.62}$$

and

$$P_i = \frac{\lambda_i}{\lambda_1 + \lambda_2 + \lambda_3}. \tag{3.63}$$

When there is no dominant single target, the three eigenvalues are comparable and the entropy is close to one. On the other hand, if only one eigenvalue is different from zero the entropy will be close to zero (Cloude 2009; Lee and Pottier 2009).

As introduced previously the eigenvalues are invariants of the coherence matrix, therefore the entropy is invariant as well. Unfortunately, the entropy alone is not sufficient to characterise completely the distribution of power among the eigenvalues, since at least two ratios are required (i.e. two real parameters). For instance, two single targets with comparable power would result in high entropy, however the target in the scene is still relatively coherent. Another invariant parameter must be introduced. This parameter is called Anisotropy:

$$A = \frac{\lambda_2 - \lambda_3}{\lambda_2 + \lambda_3}. \tag{3.64}$$

This is defined between zero and one and is small when the second and third eigenvalues are comparable. A and H contains all the polarimetric information of the eigenvalues for exception of total backscattering. Together they are a powerful tool for classification (Cloude and Pottier 1997; Ferro-Famil et al. 2002; Lee et al. 1994, 1999, 2004). Four parameters can be defined combining A and H together:

(a) $(1 - H)(1 - A)$ is high when only one single target is dominant. The single target reduces the entropy, while the second two eigenvalues are similar and they do not represent physical targets (i.e. $\lambda_2 = \lambda_3 = 0$).
(b) $H(1 - A)$ can detect random processes, since all the eigenvalues are similar. Hence, the entropy is high and the anisotropy is low as for a completely unpolarised target (i.e. $\lambda_1 = \lambda_2 = \lambda_3$).
(c) HA identifies two single scattering mechanisms with approximately the same strength. Hence the entropy will be relatively high as well as the anisotropy since the last eigenvalue is much smaller than the second (i.e. $\lambda_3 = 0$).

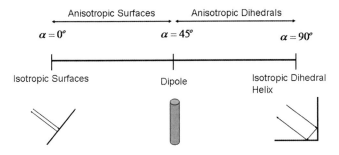

Fig. 3.4 α characteristic angle

(d) $(1 - H)A$ is high when two single targets are present but this time they have different intensity. As a consequence, the entropy is relatively low (i.e. presence of a dominant targets) but the anisotropy is high since the third eigenvalue is close to zero (i.e. $\lambda_1 \gg \lambda_2$ and $\lambda_3 = 0$).

3.4.4 α Scattering Model

Any eigenvector (or scattering mechanism in general) can be represented as:

$$\underline{u} = \left[\cos \alpha, \sin \alpha \cos \beta \, e^{j\varepsilon}, \sin \alpha \sin \beta \, e^{j\eta}\right]^T, \tag{3.65}$$

where α is the characteristic angle, β is related to the orientation angle of the target (in particular $\psi = 2\beta$) and ε, η are phase angles for the second and third components (Cloude 2009; Lee and Pottier 2009; Papathanassiou 1999). As for the Huynen parameterisation, the scattering mechanism is characterised by four parameters.

The α model has an immediate algebraic interpretation (in actual fact it was first designed as an algebraic transformation). The vector $\underline{u}$ spans all the space of targets since it can be decomposed into two rotations in a spherical coordinate system (i.e. α and β) and two changes of phase to adjust the phases of the rotated vector.

Regarding the physical interpretation, the characteristic angle α keeps information about physical properties of the target. It ranges in $\alpha \in \left[0, \frac{\pi}{2}\right]$, where the extremes are reached by isotropic targets (i.e. intermediate values are for anisotropic targets). $\alpha = 0$ represents surfaces or spheres (previously defined oddbounces), and $\alpha = \frac{\pi}{2}$ are dihedral (i.e. even bounce). $\alpha = \frac{\pi}{4}$ has the maximum anisotropic behaviour representing dipoles (in fact, the scattering matrix is of rank one and a rotation around the LOS can be considered which concentrates all the power in one linear co-polarisation). Figure 3.4 depicts the association of α with some standard targets. Please note, in order to represent a real target the phase angles must be constrained as well (Cloude 2009; Lee and Pottier 2009).

The eigenvectors obtained by the diagonalisation of the coherence matrix can be represented with the α model. $[T]$ can be written as:

$$[T] = [U] \begin{bmatrix} \lambda_1 & 0 & 0 \\ 0 & \lambda_2 & 0 \\ 0 & 0 & \lambda_3 \end{bmatrix} [U]^{*T}, \tag{3.66}$$

and $[U]$ is the unitary matrix of the eigenvectors employing the α model:

$$[U] = \begin{bmatrix} \cos\alpha_1 & \cos\alpha_2 & \cos\alpha_3 \\ \sin\alpha_1\cos\beta_1\,e^{j\varepsilon_1} & \sin\alpha_2\cos\beta_2\,e^{j\varepsilon_2} & \sin\alpha_3\cos\beta_3\,e^{j\varepsilon_3} \\ \sin\alpha_1\sin\beta_1\,e^{j\eta_1} & \sin\alpha_2\sin\beta_2\,e^{j\eta_2} & \sin\alpha_3\sin\beta_3\,e^{j\eta_3} \end{bmatrix}. \tag{3.67}$$

When the entropy is particularly low, the first eigenvector is sufficient to describe the observed target, which is approximately single. In the other cases, averaged information is required. The idea is to estimate an averaged vector able to represent the partial target. The components are obtained modelling a Bernoulli process with independent and identical distributed variable (i.e. the parameters will be averaged with the weight of their probabilities):

$$\bar{\alpha} = \sum_{i=1}^{3} P_i\alpha_i, \quad \bar{\beta} = \sum_{i=1}^{3} P_i\beta_i, \tag{3.68}$$

$$\bar{\varepsilon} = \sum_{i=1}^{3} P_i\varepsilon_i, \quad \bar{\eta} = \sum_{i=1}^{3} P_i\eta_i,$$

$$\underline{u} = \left[\cos\bar{\alpha}, \sin\bar{\alpha}\cos\bar{\beta}\,e^{j\bar{\varepsilon}}, \sin\bar{\alpha}\sin\bar{\beta}\,e^{j\bar{\eta}}\right]^T,$$

where P_i is the probability of the corresponding eigenvalue. Equation 3.67 gives averaged information about the partial target and in the case of sufficiently low entropy it characterises the physical properties of the target (Cloude and Pottier 1996).

3.5 Polarimetric Detection

The detector developed in this thesis exploits physical rather than statistical properties of the target, and does not require statistical a priori information. In the literature several polarimetric detector have been proposed (Cloude et al. 2004; De Grandi et al. 2007; Margarit et al. 2007; Novak et al. 1993a, b, 1997, 1999; Novak and Hesse 1993; Chaney et al. 1990). In this section, a brief overview of some of these detectors is provided. Both typologies of detectors with and without a priori information will be presented. However, a thorough list of statistical detectors would be exceedingly long and is beyond the scope of this thesis, since the proposed detector is focused on the physics of the scattering. Therefore, only a few significant cases will be presented. The last section is dedicated to the Polarimetric Whitening Filter. This was demonstrated to be the optimal processing for speckle reduction and it does not require statistical a priori information about the target (Chaney et al. 1990; Novak et al. 1993a; Novak and Hesse 1993). For these reasons it seems to be the best comparison for our detector.

3.5.1 Detectors Based on Statistical Approaches

In the following, a list of widely used detectors is presented (Chaney et al. 1990; Novak et al. 1993a).

(a) Optimal Polarimetric Detector (OPD):

This is a simple likelihood-ratio-test which considers the complete knowledge about the statistics of clutter and target. It can be expressed as:

$$\underline{X}^{*T}\left[\Sigma_c^{-1}\right]\underline{X} - \left(\underline{X} - \overline{\underline{X}}_t\right)^{*T}\left([\Sigma_t] + [\Sigma_c]\right)^{-1}\left(\underline{X} - \overline{\underline{X}}_t\right) > T, \qquad (3.69)$$

where $\overline{\underline{X}}_t$ is the target mean, Σ_t and Σ_c are the polarimetric covariance matrices for target and clutter respectively, and T is the detector threshold. Please note, the detector requires a priori information about the mean and covariance matrix of target and clutter. These must be adjusted to the different scenarios before any detection (Novak et al. 1987).

(b) Identity likelihood-ratio-test (ILR):

This is a variant of the OPD, where the target covariance matrix is substituted with a scaled identity matrix. Furthermore, it assumes $\overline{\underline{X}}_t = 0$ (target with zero mean, i.e. non deterministic target). The resulting detector is:

$$\underline{X}^{*T}\left[[\Sigma_c]^{-1} - \left(\frac{1}{4}E(Span(\underline{X}_t))[I] + [\Sigma_c]\right)^{-1}\right]\underline{X} > T. \qquad (3.70)$$

The algorithm still requires a priori knowledge of clutter covariance matrix plus the ratio between target and clutter (DeGraff 1988).

3.5.2 Detector Based on Physical Approaches

In this section, we will illustrate some detectors which do not use a statistical approach (at least in the first stage). However, in order to improve the detection performances they often exploit a subsequent statistical step (which will not be presented here).

(a) Single Channel Detector:

This is the simplest detector and it makes the assumption that the target to detect has a significant cross section (or at least higher than the surrounding clutter). They are based on the idea that artificial targets are mainly composed of corners and mirrors with a consequent bright backscattering. In the case where only one polarisation is accessible and a priori information are not available, a linear co-polarisation (horizontal or vertical) seems to be the best choice for detection of odd-bounces and horizontal even-bounces. The detector can be summarised as

$$\left\langle |HH|^2 \right\rangle > T. \tag{3.71}$$

The average is necessary to reduce the speckle variation (Oliver and Quegan 1998). For instance, the single pixel intensity is affected by a large statistical variation. On the other hand, with averaging, the distribution becomes closer to the mean power backscattered (i.e. smaller variance).

In some scenarios, linear co-polarisations are not the best choice, since the clutter is particularly bright with them. If dual polarimetric data are available (only one column of the scattering matrix), the cross polarisation can be exploited as well. A classical application is ship detection, where rough sea can have bright backscattering in HH and VV but not in HV. Hence, the detector would be:

$$\left\langle |HV|^2 \right\rangle > T. \tag{3.72}$$

Note that in the latter situation, we are adding physical a priori information (but not statistical).

The benefit of adopting a one polarisation approach is the relatively low complexity of the acquisition system. The drawback is that the performance is rather poor in terms of missed detections and false alarms. Missed detection can occur when the polarisation selected is a Co-pol Null for the target. As shown previously, any single target has a Co-pol Null, and if we are unfortunate to have only that polarisation available the target will be completely transparent (Kennaugh and Sloan 1952). For example, if we want to detect a vertical wire and the HH polarisation is chosen, the target backscatter will likely be below the clutter return. Evidently, another cause of missed detection is the insufficient brightness of the targets (i.e. small cross section). Depending on the target typology, this can be a significant limitation. Concerning false alarms, many natural targets have bright backscattering, constituting false alarms (e.g. an area in layover) (Woodhouse 2006).

Considering the simplicity of the detector a subsequent statistical step (sometime employing a priori information) is generally performed, in order to improve the poor performance of the physical detector alone (Kay 1998).

(b) Span Detector:

The idea is to reduce the rate of missed detection due to unfortunate choice of the antenna polarisation (equal or close to a Co-pol Null of the target) by considering the total power acquired in the scattering matrix. In other words, the entire scattering matrix $[S]$ must be acquired (i.e. quad polarimetric data) and its span is exploited:

$$Span([S]) = \left\langle |HH|^2 \right\rangle + 2\left\langle |HV|^2 \right\rangle + \left\langle |VV|^2 \right\rangle > T. \tag{3.73}$$

As in the previous case, statistical a priori information is not necessary for the execution of the algorithm. Unfortunately, we still have problems with missed detection of weak targets and false alarms from natural targets. However, better performance than the single polarisation is expected. Again a subsequent statistical step can improve the performances of the detector.

(c) Power Maximisation Synthesis (PMS):

This detector was developed as an improvement of the span detector and can be expressed as

$$\frac{1}{2}\left[\left\langle |HH|^2 \right\rangle + 2\left\langle |HV|^2 \right\rangle + \left\langle |VV|^2 \right\rangle. \right.$$

$$\left. + \sqrt{\left(\left\langle |HH|^2 \right\rangle - \left\langle |VV|^2 \right\rangle \right)^2 + 4|\langle HH^* \cdot HV \rangle + \langle HV^* \cdot VV \rangle|^2} \right] > T.$$

Again quad polarimetric data are needed and it does not use statistical a priori information (Boerner et al. 1988; DeGraff 1988).

3.5.3 *Polarimetric Whitening Filter*

This technique (first introduced by Novak) constitutes a processing strategy able to reduce optimally the speckle (Novak et al. 1993a; Novak and Hesse 1993). This method is separated from the other two categories since it is an algorithm based on statistical signal processing of quad polarimetric data but it does not employ any a priori statistical information. Moreover it was demonstrated to have the best performance among algorithms without a priori information. Considering that it seems to be the best candidate for comparison with our detector, we decided to provide a more extended description.

Polarimetric Whitening Filter (PWF) is a technique able to reduce the standard deviation of the backscattering intensity associated with the speckle effect. It assumes that the speckle-reduced pixels have a quadratic form with

$$w = \underline{u}^{*T}[A]\underline{u} \tag{3.75}$$

where $[A]$ is a Hermitian positive definite matrix and $\underline{u}$ is a generic scattering vector. The matrix $[A]$ is chosen to minimise the ratio s/m with s the standard deviation and m the mean of the intensity. We introduce the polarimetric coherence matrix $[C]$.

The matrix $[B] = [C][A]$ is still Hermitian (since product of two Hermitian matrix) and its eigenvalues, λ_1, λ_2 and λ_3 can be extracted. PWF wants to minimise

$$\left(\frac{s}{m}\right) = \frac{\sqrt{VAR[w]}}{E[w]}. \tag{3.76}$$

It can be demonstrated that

$$E[w] = Trace([B]) = \sum_{i=1}^{3} \lambda_i \quad \text{and}$$

$$VAR[w] = Trace([B])^2 = \sum_{i=1}^{3} \lambda_i^2 \tag{3.77}$$

The minimisation of Eq. 3.76 is equivalent to the minimisation of

$$\frac{\sum_{i=1}^{3} \lambda_i^2}{\left(\sum_{i=1}^{3} \lambda_i\right)^2}. \tag{3.78}$$

Using the Lagrangian multiplier β, the minimum solution is:

$$\beta = \frac{\lambda_1}{\left(\sum_{i=1}^{3} \lambda_i\right)^2} = \frac{\lambda_2}{\left(\sum_{i=1}^{3} \lambda_i\right)^2} = \frac{\lambda_3}{\left(\sum_{i=1}^{3} \lambda_i\right)^2}. \tag{3.79}$$

Hence

$$\lambda_1 = \lambda_2 = \lambda_3 = \lambda. \tag{3.80}$$

If the eigenproblem for $[B]$ is explicated for the minimum solution we have

$$[C][A] = \lambda[I] \tag{3.81}$$

Finally, in order to have the minimisation of Eq. 3.76 the matrix $[A]$ must be chosen as

$$[A] = \lambda[C]^{-1} \tag{3.82}$$

In order to obtain equal diagonal terms in the covariance matrix, a change of basis can be performed. Over the new axis selected, the power will be equally distributed. This is the reason for the name Whitening Filter, since it makes the clutter look "white".

If the hypothesis of reflection-symmetry is performed the new basis is:

$$\left[HH, \frac{HV}{\sqrt{\varepsilon}}, \frac{VV - \rho^* \sqrt{\gamma} HH}{\sqrt{\gamma\left(1 - |\rho|^2\right)}} \right],$$

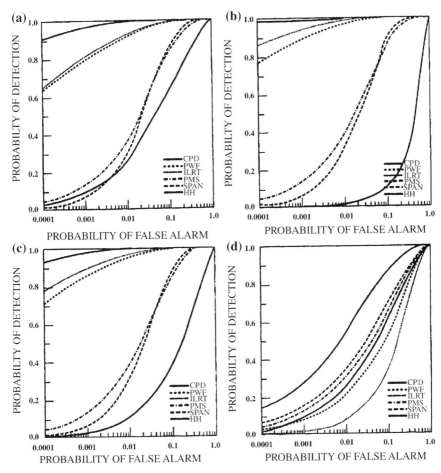

Fig. 3.5 *ROC* comparison among several detector. *OPD* Optimal Polarimetric Detector, *PWF* Polarimetric Whitening Filter, *ILRT* Identity Likelihood-Ratio-Test, *PMS* Power Maximisation Synthesis (Chaney et al. 1990). **a** Detection performance for a dihedral oriented at 0° with a 3 dB T/C ratio. **b** Detection performance for a dihedral oriented at 22.5° with a 3 dB T/C ratio. **c** Detection performance for a dihedral oriented at 45° with a 3 dB T/C ratio. **d** Detection performance for a trihedral 3 dB T/C ratio

$$\varepsilon = \frac{E\left[|HV|^2\right]}{E\left[|HH|^2\right]}, \quad \gamma = \frac{E\left[|VV|^2\right]}{E\left[|HH|^2\right]} \quad \text{and} \quad \rho = \frac{E[HH \cdot VV^*]}{\sqrt{E\left[|HH|^2\right]E\left[|VV|^2\right]}}. \quad (3.83)$$

The image obtained with the PWF i_{PWF} has an optimal speckle reduction. Subsequently, the detection can be accomplished by setting a threshold on the image intensity (Novak et al. 1993a).

$$|i_{PWF}| = |HH|^2 + \frac{|HV|^2}{\varepsilon} + \frac{\left|VV - \rho^*\sqrt{\gamma}HH\right|^2}{\gamma\left(1 - |\rho|^2\right)}$$

$$|i_{PWF}| > T. \tag{3.84}$$

The PWF will be tested in the validation chapter as the key competing method to the proposed new polarimetric detector. A complete analysis of the PWF performance is presented in that chapter, here we just mention the problems of missed detection due to weak targets and targets with partially developed speckle (e.g. targets under foliage).

3.5.4 Comparison of Detectors

Figure 3.4 shows a comparison of detectors Receiving Operative Characteristic (ROC) performed by Chaney for two different targets, a dihedral with different orientations and a trihedral (Chaney et al. 1990).

The best performances is achieved with the Optimum Polarimetric Detector (OPD), since it exploits a priori information of clutter and target. Clearly, the more the detector is provided with additional information, the closer it is to the ideal case (deterministic detector). However, the PWF is reasonably close to the OPD results. Moreover, PWF shows the best performance compared to other detectors without a priori information. The plots obtained in Fig. 3.4 will be used in the following to perform a theoretical comparison with the proposed detector (Chap. 5) (Fig. 3.5).

References

Agrawal AB, Boerner WM (1989) Redevelopment of Kennaugh's target characteristic polarization state theory using the polarization transformation ratio formalism for the coherent case. IEEE Trans Geosci Remote Sens 27:2–14

Azzam RMA, Bashara NM (1977) Ellipsometry and polarized light. North Holland Press, Amsterdam

Barakat R (1981) Bilinear constraints between the elements of the 4 × 4 Mueller-Jones matrix of polarization theory. Opt Commun 38:159–161

Bebbington DH (1992) Target vectors: spinorial concepts. Proceedings of the 2nd international workshop on radar polarimetry, IRESTE, Nantes, France, pp 26–36

Beckmann P (1968) The depolarization of electromagnetic waves. The Golem Press, Boulder

Boerner W-M (1981) Use of polarization in electromagnetic inverse scattering. Radio Sci 16:1037–1045

Boerner WM (2004) Basics of radar polarimetry. RTO SET Lecture Series

Boerner WM, El-Arini MB, Chan CY, Mastoris PM (1981) Polarization dependence in electromagnetic inverse problems. IEEE Trans Antennas and Propag 29:262–271

Boerner WM, Kostinski A, James B (1988) On the concept of the polarimetric matched filter in high resolution radar imagery: an alternative for speckle reduction. In: Proceedings of IGARSS '88 symposium, Edinburgh, Scotland, pp 69–72

Boerner WM, Yan WL, Xi AQ, Yamaguchi Y (1991) The characteristic polarization states for the coherent and partially polarized case. Proc IEEE Antennas Propag Conf, ICAP 79:1538–1550

Boerner WM, Mott H, Luneburg E (1997) Polarimetry in remote sensing: basic and applied concepts. In: IEEE Proceedings on geosciences and remote sensing symposium IGARSS, vol 3, pp 1401–1403 August 1997

Born M, Wolf E (1965) Principles of optics, 3rd edn. Pergamon Press, New York

Chaney RD, Bud MC, Novak LM (1990) On the performance of polarimetric target detection algorithms. Aerosp Electron Syst Mag IEEE 5:10–15

Cloude SR (1986) Group theory and polarization algebra. OPTIK 75:26–36

Cloude SR (1987) Polarimetry: the characterisation of polarisation effects in EM scattering. Electronics Engineering Department. York, University of York

Cloude RS (1992) Uniqueness of target decomposition theorems in radar polarimetry. Direct Inverse Methods Radar Polarim 2:267–296

Cloude RS (1995a) An introduction to wave propagation antennas. UCL Press, London

Cloude SR (1995b) Lie groups in EM wave propagation and scattering. In: Baum C, Kritikos HN (eds) Chapter 2 in electromagnetic symmetry. Taylor and Francis, Washington, pp 91–142. ISBN 1-56032-321-3

Cloude SR (1995c) Symmetry, zero correlations and target decomposition theorems. In: Proceedings of 3rd international workshop on radar polarimetry (JIPR '95), IRESTE, pp 58–68

Cloude SR (2009) Polarisation: applications in remote sensing. Oxford University Press, New York 978-0-19-956973-1

Cloude SR, Papathanassiou KP (1998) Polarimetric SAR interferometry. IEEE Trans Geosci Remote Sens 36:1551–1565

Cloude SR, Pottier E (1996) A review of target decomposition theorems in radar polarimetry. IEEE Trans Geosci Remote Sens 34:498–518

Cloude SR, Pottier E (1997) An entropy based classification scheme for land applications of polarimetric SAR. IEEE Trans Geosci Remote Sens 35:68–78

Cloude SR, Corr DG, Williams ML (2004) Target detection beneath foliage using polarimetric synthetic aperture radar interferometry. Waves Random Complex Media 14:393–414

Collin R (1985) Antennas and radiowave propagation. Mcgraw Hill, New York

De Grandi GD, Lee J-S, Schuler DL (2007) Target detection and texture segmentation in polarimetric SAR images using a wavelet frame: theoretical aspects. IEEE Tran Geosci Remote Sens 45:3437–3453

Degraff SR (1988) SAR image enhancement via adaptive polarization synthesis and polarimetric detection performance. Polarimetric Technology Workshop, Redstone Arsenal, AL

Deschamps GA (1951) Geometrical representation of the polarization of a plane electromagnetic wave. Proc IRE 39:540–544

Deschamps GA, Edward P (1973) Poincare sphere representation of partially polarized fields. IEEE Trans Antennas Propag 21:474–478

Dong Y, Forster B (1996) Understanding of partial polarization in polarimetric SAR data. Int J Remote Sens 17:2467–2475

Ferro-Famil L, Pottier E, Lee J (2002) Classification and interpretation of polarimetric SAR data. IGARSS, IEEE international geoscience and remote sensing symposium, Toronto, Canada

Freeman A (1992) SAR calibration: an overview. IEEE Trans Geosci Remote Sens 30:1107–1122

Goldstein DH, Collett E (2003) Polarized light. CRC, Boca Raton

Graves CD (1956) Radar polarization power scattering matrix. Proc IRE 44:248–252

Huynen JR (1970) Phenomenological theory of radar targets. Delft, Technical University, The Netherlands

Jones R (1941) A new calculus for the treatment of optical systems. I. description and discussion; II. Proof of the three general equivalence theorems; III. The Stokes theory of optical activity. J Opt Soc Am 31:488–503

Kay SM (1998) Fundamentals of statistical signal processing, volume 2: detection theory. Prentice Hall, Upper Saddle River

Kennaugh EM (1981) Polarization dependence of radar cross sections—a geometrical interpretation. IEEE Trans Antennas Propag 29:412–414

Kennaugh EM, Sloan RW (1952) Effects of type of polarization on echo characteristics. Ohio state University, Research Foundation Columbus, Quarterly progress reports (In lab)

Kostinski AB, Boerner W-M (1986) On foundations of radar polarimetry. IEEE Trans Antennas Propag 34:1395–1404

Krogager E (1993) Aspects of polarimetric radar imaging. Lyngby, DK, Technical University of Denmark

Lang RH (1981) Electromagnetic scattering from a sparse distribution of lossy dielectric scatterers. Radio Sci 16:15–30

Lee JS, Pottier E (2009) Polarimetric radar imaging: from basics to applications. CRC Press, Boca Raton

Lee JS, Grunes MR, Kwok R (1994) Classification of multi-look polarimetric SAR imagery based on the complex Wishart distribution. Int J Remote Sens 15:2299–2311

Lee JS, Grunes MR, Ainsworth TL, Du LJ, Schuler DL, Cloude SR (1999) Unsupervised classification using polarimetric decomposition and the complex wishart classifier. IEEE Trans Geosc Remote Sens 37:2249–2258

Lee JS, Grunes MR, Pottier E, Ferro-Famil L (2004) Unsupervised terrain classification preserving polarimetric scattering characteristics. IEEE Trans Geosci Remote Sens 42: 722–732

Lüneburg E (1995) Principles of radar polarimetry. Proc IEICE Trans Electron Theory E78-C:1339–1345

Margarit G, Mallorqui JJ, Fabregas X (2007) Single-pass polarimetric SAR interferometry for vessel classification. IEEE Trans Geosci Remote Sens 45:3494–3502

Mott H (2007) Remote sensing with polarimetric radar. Wiley, Hoboken

Novak LM, Hesse SR (1993) Optimal polarizations for radar detection and recognition of targets in clutter. In: Proceedings, IEEE national radar conference, Lynnfield, MA, pp 79–83

Novak LM, Sechtinand MB, Cardullo MJ (1987) Studies of target detection algorithms that use polarimetric radar data. In: Proceedings of the 21st Asilomar Conference on Signals, Systems andConiputers. Pacific Grove, CA

Novak LM, Burl MC, Irving MW (1993a) Optimal polarimetric processing for enhanced target detection. IEEE Trans Aerosp Electron Syst 20:234–244

Novak LM, Owirka GJ, Netishen CM (1993b) Performance of a high-resolution polarimetric SAR automatic target recognition system. Linc Lab J 6:11–24

Novak LM, Halversen SD, Owirka GJ, Hiett M (1997) Effects of polarization and resolution on SAR ATR. IEEE Trans Aerosp Electron Syst 33:102–116

Novak LM, Owirka GJ, Weaver AL (1999) Automatic target recognition using enhanced resolution SAR data. IEEE Trans Aerosp Electron Syst 35:157–175

Oliver C, Quegan S (1998) Understanding synthetic aperture radar images. Artech House, Norwood

Papathanassiou KP (1999) Polarimetric SAR interferometry. Physics, Technical University Graz

Pottier E (1992) On Dr J.R. Huynen's main contributions in the development of polarimetric radar techniques, and how the radar targets phenomenological concept becomes a theory. SPIE Opt Eng 1748:72–85

Rothwell EJ, Cloud MJ (2001) Electromagnetics. CRC Press, Boca Raton

Sinclair G (1950) The transmission and reception of elliptically polarized waves. Proc IRE 38:148–151

Strang G (1988) Linear algebra and its applications, 3rd edn. Thomson Learning, New York

Stratton JA (1941) Electromagnetic theory. McGraw-Hill, New York

Touzi R, Boerner WM, Lee JS, Lueneburg E (2004) A review of polarimetry in the context of synthetic aperture radar: concepts and information extraction. Can J Remote Sens 30:380–407

Ulaby FT, Elachi C (1990) Radar polarimetry for geo-science applications. Artech House, Norwood

Van der Mee CVM, Hovenier JW (1992) Structure of matrices transforming stokes parameters. J Math Phys 33:3574–3584

Van Zyl JJ (1992) Application of cloude's target decomposition theorem to polarimetric imaging radar data. SPIE Proc 1748:23–24

Van Zyl J, Papas C, Elachi C (1987a) On the optimum polarizations of incoherently reflected wave. IEEE Trans Antennas Propag AP-35:818–825

Van Zyl JJ, Zebker H, Elachi C (1987b) Imaging radar polarization signatures: theory and observation. Radio Sci 22:529–543

Wolf E (2003) Unified theory of coherence and polarization of random electromagnetic beams. Phys Lett 312:263–267

Woodhouse IH (2006) Introduction to microwave remote sensing. CRC Press Taylor & Francis Group, Boca Raton

Xi A-Q, Boerner WM (1992) Determination of the characteristic polarization States of the target scattering matrix [S(AB)] for the coherent monostatic and reciprocal propagation space using the polarization transformation ratio formulation. J Opt Soc Am 9:437–455

Zebker HA, Van Zyl JJ (1991) Imaging radar polarimetry: a review. Proc IEEE 79:1583–1606

Chapter 4
Polarimetric Detector

4.1 Introduction

After the introduction to polarimetry provided in the previous chapter we are now ready to develop the new polarimetric detector. This was already published and presented in international conferences: (Marino et al. 2009, 2010, in press; Marino and Woodhouse 2009). In this chapter, two different derivation approaches will be followed: the first is associated with an algebraic manipulation while the second follows a target physics operation. We believe that in this way a larger picture of the detector will be provided. As a subsequent step, the derived mathematical expression will be optimised with the purpose of removing eventual biases and improving the performance.

4.2 Derivation Using an Algebraic Approach

In this section, the detector was developed starting from the algebraic representation of a single target and implementing some manipulations through a polarimetric coherence. The algebraic point of view is sought, with the purpose of obtaining a clearer and mathematically more elegant formulation. On the other hand, next section will deal with a derivation, which takes into account the physical process of the detector.

4.2.1 Weighting of Vector Components

Any single target can be represented with a complex vector in a three dimensional space. Specifically, the algebra is constructed on a three dimensional special

A. Marino, *A New Target Detector Based on Geometrical Perturbation Filters for Polarimetric Synthetic Aperture Radar (POL-SAR)*, Springer Theses, DOI: 10.1007/978-3-642-27163-2_4, © Springer-Verlag Berlin Heidelberg 2012

unitary group $SU(3)$, applied on the field of complex numbers (Cloude 1986; 1995a; Bebbington 1992).

Given a vector $\underline{x}$ in such space, it is always possible to write the identity:

$$[I]\underline{x} = \underline{x}. \tag{4.1}$$

where $[I]$ is the identity matrix. Equation 4.1 is a special case of the general transformations

$$[A]\underline{x} = \underline{b}. \tag{4.2}$$

In our case $[A]$ is a square 3×3 matrix, but in general can be any $N \times 3$ matrix (this expression represents a linear system), since $\underline{x}$ is a three dimensional column vector. $[A]$ is a transformation of the vector $\underline{x}$ in a resulting vector $\underline{b}$, which lies in another subspace (Strang 1988, Hamilton 1989; Rose 2002). Two such subspaces exist:

(a) The subspace spanned by the column of $[A]$, also named the column subspace.
(b) The null space, which is the orthogonal complement to the column subspace.

In the case $[A]$ is a matrix of full rank, the column space is the entire $\mathbb{C}^3$ and the null space contains solely the null vector $\underline{0}$. Now, if $[A]$ is a diagonal matrix the columns of $[A]$ will always represent a basis for the entire $\mathbb{C}^3$ space (as long as all the elements of the diagonal are different from zero).

In particular, if $[A] = [I]$ the transformation is from the entire space to the entire space using the same ortho-normal basis. Clearly, this transformation leads to $\underline{b} = \underline{x}$. In the case $[A]$ is a diagonal matrix with at least one element different from 1, the transformation space for the resulting vector is still $\mathbb{C}^3$, but the basis used is not the coordinate one (the axes are not normalised vectors).

The matrix $[A]$ can be formed as:

$$[A] = \begin{bmatrix} a_1 & 0 & 0 \\ 0 & a_2 & 0 \\ 0 & 0 & a_3 \end{bmatrix}. \tag{4.3}$$

with a_1, a_2 and a_3 complex numbers. In the following, to express a diagonal matrix we will use the formalism: $[A] = diag(a_1, a_2, a_3)$. Furthermore, we identify the column vectors of $[A]$ as $\underline{a}_1 = [a_1, 0, 0]^T$, $\underline{a}_2 = [0, a_2, 0]^T$ and $\underline{a}_3 = [0, 0, a_3]^T$. The column basis is orthogonal but not ortho-normal since the basis vectors are not unitary.

By definition of basis, any vector $\underline{x}$ in the space can be expressed by the linear combination of the basis elements (i.e. the columns of $[A]$). Hence,

$$\underline{b} = x_1\underline{a}_1 + x_2\underline{a}_2 + x_3\underline{a}_3. \tag{4.4}$$

If the coordinate basis is defined as: $\underline{e}_1 = [1, 0, 0]^T$, $\underline{e}_2 = [0, 1, 0]^T$ and $\underline{e}_3 = [0, 0, 1]^T$, the linear combination in Eq. 4.4 can be rewritten as

$$\underline{b} = a_1 x_1 \underline{e}_1 + a_2 x_2 \underline{e}_2 + a_3 x_3 \underline{e}_3. \tag{4.5}$$

Therefore, the transformation $[A]\underline{x} = \underline{b}$ can be seen as a weighting of the $\underline{x}$ components for the elements on the diagonal of $[A]$. This weighting clearly will redefine the entire metric of the space, where all the vectors will be stretched along a preferential axis (Strang 1988).

4.2.2 Detector

The first step of the detector is still the definition of the vectors $\underline{\omega}_T$ and $\underline{\omega}_P$. In order to keep the development exclusively algebraic, the perturbation process could be achieved using the α parameterisation where the parameters are interpreted as rotations and phase changes (please note, the α model was born as an algebraic operation on scattering mechanisms) (Cloude 2009, Cloude and Pottier 1996). Subsequently, a change of basis that performs $\underline{\omega}_T = [1, 0, 0]^T$ is applied.

The standard Euclidean inner product between $\underline{\omega}_T$ and $\underline{\omega}_P$ can be written as $\underline{\omega}_T^{*T}\underline{\omega}_P$ (Hamilton 1989). It is always possible to consider the identities $[I]\underline{\omega}_T = \underline{\omega}_T$ and $[I]\underline{\omega}_P = \underline{\omega}_P$ consequently the inner product can be rewritten as

$$\left([I]\underline{\omega}_T\right)^{*T}\left([I]\underline{\omega}_P\right) = \left(\underline{\omega}_T^{*T}[I]\right)\left([I]\underline{\omega}_P\right) = \underline{\omega}_T^{*T}[I]\underline{\omega}_P = \underline{\omega}_T^{*T}\underline{\omega}_P. \tag{4.6}$$

In the previous section, a procedure to achieve weighting of the components is described (i.e. multiplication for a diagonal matrix). The weighting of the scattering mechanism components can be accomplished with

$$[A]\underline{\omega}_T = \underline{b}_T \quad \text{and} \quad [A]\underline{\omega}_P = \underline{b}_P. \tag{4.7}$$

We can define the weighted inner product as $\underline{b}_T^{*T}\underline{b}_P$

$$\left([A]\underline{\omega}_T\right)^{*T}\left([A]\underline{\omega}_P\right) = \underline{\omega}_T^{*T}\left([A]^{*T}[A]\right)\underline{\omega}_P = \underline{\omega}_T^{*T}[\widehat{P}]\underline{\omega}_P. \tag{4.8}$$

The operation sets a preferential direction in the targets complex space which is correspondent to the target actually present in the data. Practically,

$$[A] = diag(k_1, k_2, k_3), \tag{4.9}$$

where $\underline{k} = [k_1, k_2, k_3]^T$.

At this point, a clarification on Eq. 4.9 is required. The inner product cannot be calculated pixel by pixel, since the pixel statistical variation (i.e. speckle) can result in improper estimation of the actual target (Lee 1986; López-Martínez and Fàbregas 2003; Oliver and Quegan 1998; Touzi et al. 1999). The average over independent realisations is essential to obtain reliable results. For this reason, the instantaneous inner product $\underline{b}_T^{*T}\underline{b}_P$ is substituted with the averaged one $\langle\underline{b}_T^{*T}\underline{b}_P\rangle$:

$$\left\langle\left([A]\underline{\omega}_T\right)^{*T}\left([A]\underline{\omega}_P\right)\right\rangle = \underline{\omega}_T^{*T}\langle[A]^{*T}[A]\rangle\underline{\omega}_P = \underline{\omega}_T^{*T}[P]\underline{\omega}_P, \tag{4.10}$$

$$[P] = diag\left(\left\langle |k_1|^2\right\rangle, \left\langle |k_2|^2\right\rangle, \left\langle |k_3|^2\right\rangle\right). \tag{4.11}$$

Because $[A]$ is a diagonal matrix, $[P]$ will be diagonal as well and its terms are the averaged squared amplitude of the complex numbers on the diagonal of $[A]$, hence $[P]$ is positive definite. Please note, the expression of $[P]$ is exactly equivalent to the one obtained in the previous section after neglecting the cross terms.

The last step is the normalisation of the weighted inner product:

$$\gamma_d = \frac{\left|\underline{\omega}_T^{*T}[P]\underline{\omega}_P\right|}{\sqrt{\left(\underline{\omega}_T^{*T}[P]\underline{\omega}_T\right)\left(\underline{\omega}_P^{*T}[P]\underline{\omega}_P\right)}}. \tag{4.12}$$

Equation 4.12 represents the same formal expression of the detector. Aim of the next section is to obtain the same expression of Eq. 4.12 through a physical approach.

4.3 Mathematical Derivation with a Physical Approach

4.3.1 Perturbation Analysis and Coherence Detector

Any (normalized) single target can be uniquely represented in the target space with a three dimensional complex vector (Cloude 1986, 1987, 1995a, 1995b). In the previous chapter this vector was introduced as the scattering mechanism $\underline{\omega}$. Once a target (i.e. scattering mechanism) is selected its backscattering can be determined as

$$i(\underline{\omega}) = \underline{\omega}^{*T}\underline{k}. \tag{4.13}$$

From an algebraic point of view, Eq. 4.13 represents the inner product between the scattering vector $\underline{k}$ and the scattering mechanism $\underline{\omega}$. Additionally, the operation can be interpreted as the projection of the observables (i.e. $\underline{k}$) on the selected target (i.e. $\underline{\omega}$) since any scattering mechanism is unitary (Strang 1988). If the target of interest $\underline{\omega}$ is one component of the total polarimetric return (as for target decompositions), the operation in Eq. 4.13 extracts the component of interest from the observables (Cameron and Leung 1990; Cloude and Pottier 1996; Krogager and Czyz 1995). $i(\underline{\omega})$ is a complex number representing the pixel of a single look complex (SLC) image displaying the backscattering from the target of interest.

When two scattering mechanisms, $\underline{\omega}_1$ and $\underline{\omega}_2$ are selected (i.e. two single targets), two different images can be extracted from the observables $i(\underline{\omega}_1)$ and $i(\underline{\omega}_2)$. In the previous chapter, the polarimetric coherence was defined as

$$\gamma = \frac{\left\langle i(\underline{\omega}_1)\, i(\underline{\omega}_2)^*\right\rangle}{\sqrt{\left\langle i(\underline{\omega}_1)\, i(\underline{\omega}_1)^*\right\rangle\left\langle i(\underline{\omega}_2)\, i(\underline{\omega}_2)^*\right\rangle}}. \tag{4.14}$$

It estimates the correlation between the two images (Boerner 2004; Mott 2007). If these are similar, the amplitude of the polarimetric coherence γ will be close to 1. We want to demonstrate:

Given a scattering mechanism $\underline{\omega}_1$ proportional to the target to be detected, and given a second scattering mechanism $\underline{\omega}_2$ close to $\underline{\omega}_1$ within the target space, the polarimetric coherence is high if in the averaging cell the component of interest (proportional to $\underline{\omega}_1$) is stronger than the other two orthogonal components.

(1) The representation of a single target as a scattering vector is dependent on the basis selected to represent the target space (Cloude 2009). In the following demonstration we decided to use the Pauli basis as the starting point, however any other basis could be selected leading to exactly the same mathematical result. A given scattering vector in Pauli basis can be represented as $\underline{k}^P$, while the scattering mechanisms for the target of interest is $\underline{\omega}_T^P$.

The first step in the detector design is a change of basis aimed at overlapping one of the axes (of the new basis) with the target of interest $\underline{\omega}_T^P$. This is always possible since any single target is uniquely represented by a vector that can constitute one axis of the basis. The operation is achievable by multiplying by a unitary matrix $[U]$ (Lee and Pottier 2009),

$$\underline{\omega}_T = [U]\underline{\omega}_T^P = [1,\ 0,\ 0]^T. \tag{4.15}$$

In the new basis, the target of interest lies only in one component of the three dimensional complex vector (since $\underline{\omega}_T$ itself is one axis of the basis). The absolute phase does not constitute an exploitable parameter and can be set to zero without lost of generality. Following the initial arrangement $\underline{\omega}_T = \underline{\omega}_1$. The other two axes must be chosen orthogonal to $\underline{\omega}_T$ and will be regarded as $\underline{\omega}_{C2}$ and $\underline{\omega}_{C3}$ (Hamilton 1989). Therefore,

$$\underline{\omega}_T \perp \underline{\omega}_{C2} \perp \underline{\omega}_{C3}. \tag{4.16}$$

Once the new basis is selected, the scattering vector needs to be expressed in this basis,

$$\underline{k} = [U]\underline{k}^P = [k_1,\ k_2,\ k_3]^T. \tag{4.17}$$

where $k_1,\ k_2,\ k_3 \in \mathbb{C}$. Or equivalently

$$\underline{k} = k_1\underline{\omega}_T + k_2\underline{\omega}_{C2} + k_3\underline{\omega}_{C3}. \tag{4.18}$$

Finally, the coherency matrix $[C]$ is estimated starting from the obtained k. The resulting complex image when the target $\underline{\omega}_T$ is selected is

$$i(\underline{\omega}_T) = \underline{\omega}_T^{*T}\ \underline{k} = k_1. \tag{4.19}$$

In the new basis, when the projection on $\underline{\omega}_T$ is evaluated, the components of the scattering vector k_2 and k_3 are deleted completely, since by definition they are orthogonal to the direction of $\underline{\omega}_T$. Therefore, the target to detect is solely concentrated in the k_1 component. For this reason, k_2 and k_3 can be regarded as *clutter*. Please note, the distinction between target and clutter components can be accomplished exclusively in the new basis (for instance in Pauli basis the target of interest generally does not lie in only one component). The scalar projection in Eq. 4.19 can be interpreted as an ideal filter for the target of interest, which in general will be different from zero. However, in most cases it represents the target of interest only when the other two components of the scattering vector are nearly absent. Some detectors are based on thresholds on the amplitude of the projection. Unfortunately, these detectors have two major problems:

(1.a) First, when the operation is accomplished without averaging over several representations (i.e. neighbour pixels) the results can be strongly affected by speckle (Lee 1986, López-Martínez and Fàbregas 2003, Oliver and Quegan 1998). The false alarm rate due to the surrounding partial targets will be unsuitably high. However, the straightforward remedy is to average over neighbour pixels before considering the threshold.

(1.b) Secondly, the occurrence of a strong component k_1 does not generally assure the presence of the target to detect, since different single or partial target can have significant projections on the target of interest (Cloude and Pottier 1996). In other words, the ratios between components must be considered.

(2) In the second step, we need to generate a second scattering mechanism $\underline{\omega}_2$ similar to $\underline{\omega}_T$ in the target space. This new vector will be regarded as "perturbed target", $\underline{\omega}_P$ (i.e. $\underline{\omega}_2 = \underline{\omega}_P$). Several approaches can be adopted to obtain $\underline{\omega}_P$ starting from $\underline{\omega}_T$. In the following, two of them are listed:

(2.a) Geometrical: random noise
A random vector (for instance Gaussian) with zero mean called $\underline{d\omega}$ is generated much smaller than $\underline{\omega}_T$. For instance, we could choose $\|\underline{d\omega}\| = 0.1$ (please note, the scattering mechanisms are unitary). The perturbed target is given by

$$\underline{\omega}_P = \frac{\underline{\omega}_T + \underline{d\omega}}{\|\underline{\omega}_T + \underline{d\omega}\|} \qquad (4.20)$$

(2.b) Physical: Huynen Polarisation Fork
The previous methodology is a geometrical rather than physical operation. Fortunately, the physical feasibility of the obtained vector $\underline{\omega}_P$ is assured by the completeness of the vector space (any unitary three dimensional complex vector is a physical feasible scattering mechanism) (Bebbington 1992;

Cloude 1986, 1995a). However, the perturbation can be directly related to physical changes in the target. For this reason, we want to perform the perturbation of $\underline{\omega}_T$ with a more physical approach, utilising a target parameterisation. One idea could be to move the entire polarisation fork slightly (rotating the characteristic polarisations). In fact, a slightly different polarisation fork characterises a slightly different target (Boerner et al. 1981). The small rotation of the characteristic polarizations on the Poincaré sphere can be accomplished with the Huynen parameters (Huynen 1970). In other words, if ψ_m, χ_m, υ and γ are the parameters used to define the target $\underline{\omega}_T$, the perturbed target $\underline{\omega}_P$ will be obtained by substituting

$$\psi_m \pm \Delta\psi_m, \quad \chi_m \pm \Delta\chi_m, \quad \upsilon \pm \Delta\upsilon \quad \text{and} \quad \gamma \pm \Delta\gamma, \tag{4.21}$$

where $\Delta\psi_m$, $\Delta\chi_m$, $\Delta\upsilon$ and $\Delta\gamma$ are positive real numbers corresponding to a fraction (e.g. a twelfth or a tenth) of the maximum value of the respective variable. The variation can be positive or negative in order to keep the final parameter within the allowed range. In Appendix 1, we present the proof that a slight change of the Huynen parameters generates a slightly different target. Basically, this is due to the continuity of the functions employed in the Huynen representation (if the parameters move in the allowed range of values).

The scattering mechanism to detect $\underline{\omega}_T$ in the Huynen representation is (Huynen 1970)

$$[S_T] = [R(\psi_m)][T(\chi_m)][S_d(\gamma, \upsilon)][T(\chi_m)][R(-\psi_m)],$$

$$[S_d] = \begin{pmatrix} e^{i\upsilon} & 0 \\ 0 & \tan(\gamma)e^{-i\upsilon} \end{pmatrix},$$

$$[T(\tau_m)] = \begin{pmatrix} \cos\chi_m & -i\sin\chi_m \\ -i\sin\chi_m & \cos\chi_m \end{pmatrix}, \tag{4.22}$$

$$[R(\psi_m)] = \begin{pmatrix} \cos\psi_m & -\sin\psi_m \\ \sin\psi_m & \cos\psi_m \end{pmatrix}.$$

Hence, the perturbed target can be represented as

$$[S_P] = [R(\psi_m \pm \Delta\psi_m)][T(\chi_m \pm \Delta\chi_m)][S_d(\gamma \pm \Delta\gamma, \upsilon \pm \Delta\upsilon)][T(\chi_m \pm \Delta\chi_m)]$$

$$\times [R(-(\psi_m \pm \Delta\psi_m)]$$

$$\tag{4.23}$$

If the variation is small, then $[S_P] \approx [S_T]$.

Similarly, the rotation of the Polarisation Fork can be obtained starting from the α parameterisation as (Cloude and Pottier 1997)

$$\underline{\omega}_T = \left[\cos\alpha, \ \sin\alpha\cos\beta e^{i\varepsilon}, \ \sin\alpha\sin\beta e^{i\eta}\right]^T \tag{4.24}$$

$$\underline{\omega}_P = \begin{bmatrix} \cos(\alpha \pm \Delta\alpha) \\ \sin(\alpha \pm \Delta\alpha)\cos(\beta \pm \Delta\beta)e^{i(\varepsilon \pm \Delta\varepsilon)} \\ \sin(\alpha \pm \Delta\alpha)\sin(\beta \pm \Delta\beta)e^{i(\eta \pm \Delta\eta)} \end{bmatrix} \tag{4.25}$$

where again $\Delta\alpha$, $\Delta\beta$, $\Delta\varepsilon$ and $\Delta\eta$ are a fraction of the maximum value of the respective variables (for ε and η the maximum value is fixed to 2π). Again, $[S_P] \approx [S_T]$ and $\underline{\omega}_P \approx \underline{\omega}_T$.

The optimisation of this procedure is treated in the following sections.

At this point, a clarification concerning the basis used is needed. The Huynen parameterisation is formulated on the Lexicographic basis while the α model employs the Pauli basis. Therefore, a change of basis on $\underline{\omega}_P$ must be considered after the perturbation process.

After the change of basis, the perturbed target is a unitary three dimensional complex vector $\underline{\omega}_P = [a, \ b, \ c]^T$, with a, b and c complex numbers. Considering $\underline{\omega}_P \approx \underline{\omega}_T$, we must have

$$|a| \approx 1, \quad |b| \approx 0 \quad \text{and} \quad |c| \approx 0,$$

$$|a|^2 + |b|^2 + |c|^2 = 1. \tag{4.26}$$

(3) Once the two scattering mechanisms are defined the polarimetric coherence (in the new basis) can be estimated with

$$\gamma(\underline{\omega}_T, \ \underline{\omega}_P) = \frac{\langle i(\underline{\omega}_T) \ i^*(\underline{\omega}_P) \rangle}{\sqrt{\langle i(\underline{\omega}_T) \ i^*(\underline{\omega}_T) \rangle \langle i(\underline{\omega}_P) \ i^*(\underline{\omega}_P) \rangle}} \tag{4.27}$$

where:

$$\begin{aligned} \langle i(\underline{\omega}_T) \ i^*(\underline{\omega}_P) \rangle &= a\langle |k_1|^2 \rangle + b\langle k_1 \ k_2^* \rangle + c\langle k_1 \ k_3^* \rangle \\ \langle i(\underline{\omega}_T) \ i^*(\underline{\omega}_T) \rangle &= \langle |k_1|^2 \rangle \\ \langle i(\underline{\omega}_P) \ i^*(\underline{\omega}_P) \rangle &= |a|^2\langle |k_1|^2 \rangle + |b|^2\langle |k_2|^2 \rangle + |c|^2\langle |k_3|^2 \rangle \\ &\quad + 2\mathrm{Re}\big(ab^*\langle k_1 \ k_2^* \rangle\big) + 2\mathrm{Re}\big(ac^*\langle k_1 \ k_3^* \rangle\big) \\ &\quad + 2\mathrm{Re}\big(cb^*\langle k_3 \ k_2^* \rangle\big) \end{aligned} \tag{4.28}$$

After dividing numerator and denominator by $|a|\langle|k_1|^2\rangle$, the amplitude of the polarimetric coherence becomes:

$$|\gamma(\underline{\omega}_T, \underline{\omega}_P)| = \frac{\left|1 \cdot e^{j\phi_a} + \frac{b}{|a|}\frac{\langle k_1 k_2^*\rangle}{\langle|k_1|^2\rangle} + \frac{c}{|a|}\frac{\langle k_1 k_3^*\rangle}{\langle|k_1|^2\rangle}\right|}{\sqrt{\Lambda}} \qquad (4.29)$$

$$\Lambda = 1 + \frac{|b|^2}{|a|^2}\frac{\langle|k_2|^2\rangle}{\langle|k_1|^2\rangle} + \frac{|c|^2}{|a|^2}\frac{\langle|k_3|^2\rangle}{\langle|k_1|^2\rangle}$$
$$+ \frac{2\mathrm{Re}(ab^*\langle k_1 k_2^*\rangle) + 2\mathrm{Re}(ac^*\langle k_1 k_3^*\rangle) + 2\mathrm{Re}(cb^*\langle k_3 k_2^*\rangle)}{|a|^2\langle|k_1|^2\rangle} \qquad (4.30)$$

We refer to $(|b|/|a|)^2$ and $(|c|/|a|)^2$ as Reduction Ratios (*RedR*). The perturbed targets are chosen in order to have small *RedR*. Hence, in the sum the elements multiplied by the *RedR* are lowered. These terms are regarded as *clutter* terms and are all the elements except the ones with the sought component $\langle|k_1|^2\rangle$ alone (please note after the division the latter becomes 1). Two typologies of clutter terms can be identified:

(3.a) Cross-correlations terms: $\langle k_1 k_2^*\rangle$, $\langle k_1 k_3^*\rangle$, $\mathrm{Re}(ab^*\langle k_1 k_2^*\rangle)$, $\mathrm{Re}(ac^*\langle k_1 k_3^*\rangle)$ and $\mathrm{Re}(cb^*\langle k_3 k_2^*\rangle)$. These are generally small, since for partial targets the components of $\underline{k}$ are partially uncorrelated (Touzi et al. 1999). For two completely uncorrelated terms, the mean of the products becomes the product of the means, which are 0 since they are complex Gaussian zero mean (Oliver and Quegan 1998; Papoulis 1965):

$$E[k_1 k_2^*] = E[k_1]E[k_2^*] = 0 \qquad (4.31)$$

In practical cases, these terms are different from 0 for two reasons. Firstly, the components are not completely uncorrelated. Secondly, the ensample average is not performed over an infinite number of elements, hence there will be a residual correlation due to the insufficient number of samples (Touzi et al. 1999).

(3.b) Power terms: $\langle|k_2|^2\rangle$ and $\langle|k_3|^2\rangle$. They depend on the clutter power.

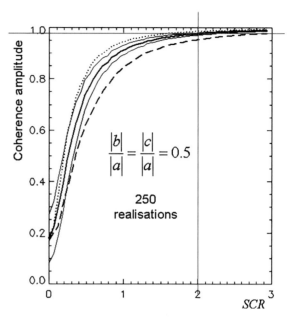

Fig. 4.1 Coherence amplitude detector: *Solid lines* mean inside the standard deviation boundaries for uncorrelated target-clutter. *Dotted line* mean in the case of positive target-clutter correlation. *Dashed line* mean in the case of negative target-clutter correlation. SCR: Signal to Clutter Ratio. Average over 250 realizations and window size 5 × 5

Finally, when $\left\langle |k_1|^2 \right\rangle$ is higher than the clutter terms, the *RedR* combined with the normalisation for $\left\langle |k_1|^2 \right\rangle$ makes the clutter terms negligible in the sum and the polarimetric coherence has amplitude close to one. If the component of interest is not dominant, the clutter terms influence the final sum more appreciably, lowering the coherence amplitude.

(4) The amplitude of the polarimetric coherence between target and perturbed target changes depending on the dominance of the target to be detected. In conclusion, the coherence amplitude can be used as a detector when a threshold is set. If T is the threshold, the detection rule can be

$$
\begin{aligned}
H_0 &: |\gamma(\underline{\omega}_T, \underline{\omega}_P)| \geq T \\
H_1 &: |\gamma(\underline{\omega}_T, \underline{\omega}_P)| < T
\end{aligned}
\tag{4.32}
$$

where H_0 is the hypothesis with target, and H_1 with only clutter (Hippenstiel 2002; Kay 1998).

With the purpose of testing the theoretical effectiveness of the detector, the simulation of the coherence amplitude estimated as a stochastic process is presented in Fig. 4.1. The simulation takes into account a deterministic target k_1 (target to be detected) and two random variables, complex Gaussian zero mean (i.e. k_2 and k_3), independent of each other (Oliver and Quegan 1998). Please note, a complete statistical assessment of the detector will be accomplished in the next chapter. Here, the performed simulation has the simple purpose of giving a visual interpretation of the detector. The plot shows the mean value of the coherence (over 250 realizations) confined in the standard deviation boundaries. A 5×5 window and $RedR = (|b|/|a|)^2 = (|c|/|a|)^2 = 0.25$ are considered. The Signal to Clutter Ratio (SCR) is defined as:

$$SCR = \frac{\langle |k_1|^2 \rangle}{\langle |k_2|^2 \rangle + \langle |k_3|^2 \rangle} \qquad (4.33)$$

While SCR due to the individual clutter component can be calculated as

$$
\begin{aligned}
SCR_2 &= \langle |k_1|^2 \rangle / \langle |k_2|^2 \rangle \\
SCR_3 &= \langle |k_1|^2 \rangle / \langle |k_3|^2 \rangle
\end{aligned}
\qquad (4.34)
$$

The plot is obtained increasing simultaneously SCR_2 and SCR_3. This lead to a resulting $SCR = \dfrac{SCR_2}{2}$.

4.3.2 Bias Removal: Final Detector

The plot in Fig. 4.1 is obtained considering the components of the scattering vector $\underline{k}$ to be independent of each other. Under this hypothesis, the cross correlation terms are very small. This is an adequate approximation for partial targets with low degrees of polarisation, but it cannot be applied to single (coherent) targets (since these in general have completely correlated components).

A counterexample is found in considering the detection of horizontal dipoles (the scattering matrix has exclusively the S_{HH} element), when the target present in the cell is a 45° dipole (Cloude 2009). In lexicographic basis, $\underline{\omega}_T$ is $\underline{\omega}_T = [1, 0, 0]^T$, where the single target on the scene can be represented as $\underline{k}_L = \kappa[1, \sqrt{2}, 1]^T$, with κ a complex number. Substituting the value of $\underline{\omega}_T$ and $\underline{k}_L$ in the polarimetric coherence (Eq. 4.29) the amplitude obtained is unitary.

In conclusion, correlation between target and clutter introduces bias in the coherence amplitude. In Fig. 4.1, the dotted and dashed lines show the case when the coherent target is correlated with the two clutter components, respectively in a constructive or destructive way. The amplitude of the correlation coefficient between target and clutter components is 0.65.

The aim of this section is to remove the bias due to the correlation between the components. Firstly, we recognise that the cross terms do not add constructive information in our specific situation. In the case of uncorrelated components they merely add noise related to the finite averaging (Touzi et al. 1999) (the substitution of the expected value $E[.]$ deletes them completely). However, for high values of coherence, the bias introduced is not appreciable. On the other hand, when the $\underline{k}$ components are correlated, they introduce bias that results in false alarms or missed detections. Consequently, the detector is improved and simplified when they are ignored. The possibility to ignore the cross term is linked with the change of basis performed which makes the components independent of each other when the target to detect is present. Appendix 2 presents the proof of the appropriateness of the operation. Regarding the uniqueness of the result, a dominant single target can be completely (and uniquely) characterised, since the power terms calculated in the detector are obtained from the projections of $\underline{k}$ on the scattering mechanisms (Cloude 1986; Rose 2002).

With the purpose of neglecting the cross terms, the polarimetric coherence is substituted with another operator working on the target power components:

$$\gamma_d(\underline{\omega}_T, \underline{\omega}_P) = \frac{\left|\underline{\omega}_T^{*T}[P]\underline{\omega}_P\right|}{\sqrt{(\underline{\omega}_T^{*T}[P]\underline{\omega}_T)\,(\underline{\omega}_P^{*T}[P]\underline{\omega}_P)}}, \tag{4.35}$$

$$[P] = \begin{bmatrix} \left\langle |k_1|^2\right\rangle & 0 & 0 \\ 0 & \left\langle |k_2|^2\right\rangle & 0 \\ 0 & 0 & \left\langle |k_3|^2\right\rangle \end{bmatrix}. \tag{4.36}$$

$$\gamma_d(\underline{\omega}_T, \underline{\omega}_P) = \frac{1}{\sqrt{1 + \dfrac{|b|^2}{|a|^2}\dfrac{\left\langle |k_2|^2\right\rangle}{\left\langle |k_1|^2\right\rangle} + \dfrac{|c|^2}{|a|^2}\dfrac{\left\langle |k_3|^2\right\rangle}{\left\langle |k_1|^2\right\rangle}}}. \tag{4.37}$$

The amplitude of the modified coherence in Eq. 4.37 is regarded as the *detector*. This is dependent only on the power of the components of $\underline{k}$ and $\underline{\omega}_P$, therefore it is a real number.

The expression obtained in Eq. 4.37 is still dependent on the basis used to express the vectors $\underline{\omega}_T$ and $\underline{\omega}_P$. Here, the target to detect overlaps the first axis, hence it is present exclusively in the k_1 component (i.e. k_2, k_3 represent the clutter). If three ortho-normal vectors are considered as $\underline{e}_1 = [1, 0, 0]^T$, $\underline{e}_2 = [0, 1, 0]^T$ and $\underline{e}_3 = [0, 0, 1]^T$, the power of target and clutter can be written as:

$$P_T = \left\langle |\underline{k}^{T*} \cdot \underline{e}_1|^2\right\rangle, \quad P_{C2} = \left\langle |\underline{k}^{T*} \cdot \underline{e}_2|^2\right\rangle \quad \text{and} \quad P_{C3} = \left\langle |\underline{k}^{T*} \cdot \underline{e}_3|^2\right\rangle. \tag{4.38}$$

Consequently, Eq. 4.37 can be rewritten as:

$$\gamma_d = \frac{1}{\sqrt{1 + \dfrac{|b|^2}{|a|^2}\dfrac{P_{C2}}{P_T} + \dfrac{|c|^2}{|a|^2}\dfrac{P_{C3}}{P_T}}}. \tag{4.39}$$

Fig. 4.2 Detector: mean over 250 realisations inside the standard deviation boundaries. Window size 5 × 5

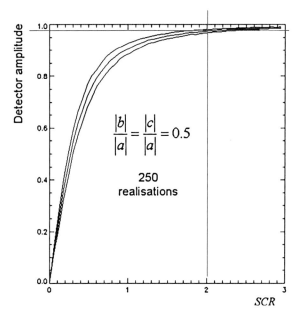

Looking at Eq. 4.39, the lowering effect played by the *RedR* is clear. If the clutter power is lower than the target power the two terms on the denominator are negligible and $\gamma_d = \dfrac{1}{\sqrt{1 + \varepsilon}} \approx 1$, with $\varepsilon \approx 0$. Conversely, if the clutter components are significant, ε is not close to 0 and the denominator is appreciably different from 1, lowering the value of the detector.

The trend of the detector can be identified in Fig. 4.2. Comparing Figs. 4.1 and 4.2 the variance appears strongly reduced for low values of coherence, moreover the two means look very close for values higher than 0.6.

The difference for lower values is related with the coherence bias due to finite averaging. The bias is introduced by the cross terms, thus it disappears when we neglect them. For high values of clutter the detector becomes close to 0. The two extremes of the detector are achieved for:

$$\lim_{\varepsilon \to 0} \gamma_d = \lim_{\varepsilon \to 0} \frac{1}{\sqrt{1 + \varepsilon}} = 1,$$

$$\lim_{\varepsilon \to \infty} \gamma_d = \lim_{\varepsilon \to \infty} \frac{1}{\sqrt{1 + \varepsilon}} = 0. \tag{4.40}$$

For uncorrelated components, the presence of cross terms results merely in a higher variance.

The detection is achieved setting a threshold on Eq. 4.37. The decision rule is similar to the previous one:

$$H_0 : \gamma_d \geq T$$
$$H_1 : \gamma_d < T$$

(4.41)

4.3.3 Generalised Detector

The algorithm presented in the previous section is based on two main hypotheses: (i) monostatic sensor (same transmitter and receiver antenna) and (ii) reciprocal medium. In this occurrence, the two cross terms of the scattering matrix are identical $S_{12} = S_{21}$ (with the exception of noise). On these assumptions, the problem can be simplified and located in a three dimensional complex space (rather than a four dimensional one) (Cloude 2009; Lee and Pottier 2009). The hypotheses can fail in two main cases:

(a) Transmitter and receiver antennas are different as in a bistatic system (Cherniakov 2008; Willis 2005).
(b) The medium is not reciprocal. An example is the ionosphere at low frequency (e.g. P-band) due to the presence of plasma which leads to the phenomenon of Faraday rotation (Cloude 2009; Freeman 1992). This effect is observed principally in satellite radar, and can be corrected to some extent.

In the case hypotheses (i) and (ii) are not fulfilled, the detector can still be built with a procedure analogous to the three dimensional case. Now, the scattering mechanism to be detected $\underline{\omega}_{T4}$ is four dimensional complex. Again, a change of basis which makes $\underline{\omega}_{T4} = [1, 0, 0, 0]^T$ is performed. However, the perturbed target cannot be generated with the Huynen parameterisations, since this is defined for monostatic scenarios. On the other hand, the additive noise approach is still fully suitable. Nevertheless, any parameterisation developed for bistatic can be exploited (Germond et al. 2000). Finally, the perturbed target can be expressed in the new basis as $\underline{\omega}_{P4} = [a, b, c, d]^T$, where

$$a, b, c, d \in \mathbb{C}, \tag{4.42}$$

$$|a| \approx 1, \quad |b| \approx 0, \quad |c| \approx 0, \quad |d| \approx 0,$$

$$|a|^2 + |b|^2 + |c|^2 + |d|^2 = 1,$$

$$a \gg b, \quad a \gg c \quad \text{and} \quad a \gg d.$$

In 4 dimensions, the scattering vector is $\underline{k}_4 = [k_1, k_2, k_3, k_4]^T$ and the covariance matrix is calculated as $[C_4] = \langle \underline{k}_4 \, \underline{k}_4^{*T} \rangle$. With the intention of removing the bias due to the cross terms, only the elements on the diagonal are utilised:

$$[P_4] = \begin{bmatrix} \langle |k_1|^2 \rangle & 0 & 0 & 0 \\ 0 & \langle |k_2|^2 \rangle & 0 & 0 \\ 0 & 0 & \langle |k_3|^2 \rangle & \\ 0 & 0 & 0 & \langle |k_4|^2 \rangle \end{bmatrix} \tag{4.43}$$

The detector is implemented as

$$\gamma_{d4} = \frac{|\omega_{T4}^{*T}[P_4]\omega_{P4}|}{\sqrt{(\omega_{T4}^{*T}[P_4]\omega_{T4})(\omega_{P4}^{*T}[P_4]\omega_{P4})}} \tag{4.44}$$

$$\gamma_{d4} = \frac{1}{\sqrt{1 + \frac{|b|^2}{|a|^2}\frac{\langle |k_2|^2 \rangle}{\langle |k_1|^2 \rangle} + \frac{|c|^2}{|a|^2}\frac{\langle |k_3|^2 \rangle}{\langle |k_1|^2 \rangle} + \frac{|d|^2}{|a|^2}\frac{\langle |k_4|^2 \rangle}{\langle |k_1|^2 \rangle}}} \tag{4.45}$$

Equation 4.45 presents the most general form of the detector, since in the case hypotheses (i) and (ii) are fulfilled, it reduces automatically to the previous formula (Eq. 4.37). In the following the three dimensional formulation will be employed, but all the results and considerations can be adapted in a rather straightforward way to the general detector.

4.4 Detector Interpretation

The aim of this section is to provide an intuitive interpretation of the algorithm, describing the algebraic and physical reasons for the detection achievability.

4.4.1 Geometrical Interpretation

The weights used in the inner product between target and perturbed target are extracted from the observables. Here, we want to address the following question: why the weighted inner product results in a detector?

From the mathematical point of view, the question finds a simple answer considering the clutter effect on the coherence denominator. However, we want to find a justification based on the vector representation in the target space. The target of interest is the first axis of the basis (i.e. $\omega_T = [1, 0, 0]^T$), while the perturbed target has all the components (i.e. $\omega_P = [a, b, c]^T$). When the standard normalised inner product between ω_T and ω_P is estimated, the correlation (which increases the value of the coherence) is introduced by the first component solely. The second and third components cannot be correlated since ω_T does not have them at all. Being specific, the amplitude of the correlation is equal to the cosine of the angle between the two vectors (since they are normalised) (Strang 1988):

$$\left| \underline{\omega}_T^{*T} \underline{\omega}_P \right| = \cos \varphi = |a|, \tag{4.46}$$

where φ is the angle between the two vectors.

However, the detector is based on a *weighed* and *normalised* inner product between $\underline{\omega}_T$ and $\underline{\omega}_P$. Since the first component is the only one bringing correlation, the inner product changes depending on the weight allocated to the first component compared to the others. Two extremes can be considered:

(a) The observed target is exactly the target of interest:

The weighting is performed with $[A] = diag(k_1, 0, 0)$. In this particular scenario, $[A]$ has rank one, hence it represents a transformation in a one-dimensional space (i.e. a complex line), the space spanned by the target to detect. Moreover, the product of any vector for the matrix $[A]$ will be the projection (plus a scaling) of the vector on this complex line (Cloude 1995b):

$$\underline{b}_P = [A]\underline{\omega}_P = k_1 a \underline{\omega}_T, \tag{4.47}$$

with a the first component of $\underline{\omega}_P$. In other words, the multiplication by $[A]$ deletes the second and third component of $\underline{\omega}_P$.

After the weighting, the coherence will be one (and the target detected), since the normalised inner product between two parallel vectors is one (i.e. $\cos 0 = 1$).

(b) The target of interest is completely absent:

In this case, $[A] = diag(0, k_2, k_3)$, the matrix has $rank([A]) = 2$ and the space of the columns represents a complex plane, which is perpendicular to the direction of the target of interest. The multiplication of a vector by $[A]$ will project (and scale) the vector on this complex plane:

$$\underline{b}_P = [A]\underline{\omega}_P = k_2 b \underline{\omega}_{C2} + k_3 c \underline{\omega}_{C3} \tag{4.48}$$

The resulting vector is a combination of the two clutter terms and consequently, it is orthogonal to the target of interest. The result of the inner product will be zero (i.e. $\cos(\pi/2) = 0$).

In an intermediate case, $[A]$ always has full rank (i.e. $\det([A]) \neq 0$) due to the polarimetrically white noise (e.g. thermal noise), which is spread in all the components. Generally, the weighting has two main effects on the scattering mechanisms: a rotation and a rescaling (Rose 2002; Strang 1988). The scaling effect can be neglected since the inner product is subsequently normalised. On the other hand the rotation has effects only on $\underline{\omega}_P$, since $\underline{\omega}_T$ cannot change direction and it will always be along the first component,

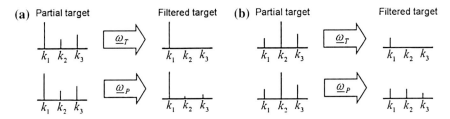

Fig. 4.3 Visual explanation of the filter with target and perturbed target. **a** Detection achieved. **b**. No detection achieved

$$\forall [A], \ \underline{b}_T \ = \ [A]\underline{\omega}_T \ = \ k_1\underline{\omega}_T. \tag{4.49}$$

Because $\underline{\omega}_T$ has only the first component, the other diagonal elements of $[A]$ have no effect on it.

In conclusion, if the rotation makes the resulting vector $\underline{b}_P$ closer to $\underline{b}_T$ the angle between them will reduce and the coherence increases. Specifically, the angle between $\underline{\omega}_T$ and $\underline{\omega}_P$ before the weighting can be calculated as

$$\varphi \ = \ \cos^{-1}(|a|). \tag{4.50}$$

The weighting on target and perturbed target works as

$$[A]\underline{\omega}_T \ = \ [k_1, \ 0, \ 0]^T = \ k_1\underline{\omega}_T,$$

$$[A]\underline{\omega}_P \ = \ [k_1a, \ k_2b, \ k_3c]^T \ = \ k_1a\underline{\omega}_T \ + \ k_2b\underline{\omega}_{C2} \ + \ k_3c\underline{\omega}_{C3}. \tag{4.51}$$

The normalised inner product between the weighted scattering mechanisms is the detector, consequently the angle between the vectors becomes $\hat{\vartheta} \ = \ \cos^{-1}(\gamma_d)$. The angle decreases after the weighting if

$$\hat{\vartheta} \ = \ \cos^{-1}(\gamma_d) \ < \ \cos^{-1}(|a|) \ = \ \varphi$$

$$\gamma_d \ > \ |a| \tag{4.52}$$

Geometrically, this is obtained when the observed target has a k_1 component stronger than the others. In other words, the correlation increases if $\underline{\omega}_P$ is stretched in a direction where the k_1 component is stronger.

The fact that the angle is reduced is not sufficient to guarantee detection, since the detector γ_d is required to be over the threshold as well.

4.4.2 Physical Interpretation

The detector can be interpreted as a filter, represented in Fig. 4.3 as a simple schematic. The vertical bars stand for the power of the scattering vector components.

After the change of basis, which makes $\underline{\omega}_T = [1, 0, 0]^T$,, k_1 represents the target to detect and k_2, k_3 are the clutter.

The final image (as interpreted by the detector) is obtained as the incoherent sum of the three components. As explained previously, the image formation (i.e. scalar projection) behaves similarly to a filter. The first row of any example (i.e. $\underline{\omega}_T$) is ideal and deletes completely the orthogonal clutter components.

The second row (i.e. $\underline{\omega}_P$) results in a linear combination of the sought component (slightly lowered) plus a small amount of the orthogonal ones. In (a) the match between the target and perturbed target is high, since the power in the two images is similar. This is not true in (b), since the $\underline{\omega}_P$ image has much more power in the orthogonal component than the $\underline{\omega}_T$ one, hence the $\underline{\omega}_P$ power significantly lowers the coherence.

4.5 Parameters Selection

The selectivity of the proposed detector is dependent on two main parameters: threshold and *RedR*. Therefore they must be carefully chosen. The aim of this section is to find a rationale for their selection.

The detector γ_d (as expressed in Eq. 4.37) is a stochastic process (Gray and Davisson 2004; Kay 1998; Oliver and Quegan 1998; Papoulis 1965). This represents a complication in the parameters settings, since we do not deal with deterministic expressions. The detector randomness is a consequence of the clutter components of the scattering vector k_2 and k_3, namely complex Gaussian random variables with zero mean (Franceschetti and Lanari 1999; Oliver and Quegan 1998). Several realisations of the same detector are generally different depending on the statistical variability (or variance). In order to take into account the detector variation, a statistical characterisation of the detector is required, in particular, the Probability Density Function (*pdf*) must be derived (Papoulis 1965). The next chapter deals with the analytical calculation of the *pdf*, while here an easier expression is investigated which is independent of the statistical realisation. With such an expression, we can achieve an easier and more direct insight into the role played by the detector parameters.

4.5.1 Reduction Ratios (RedR) and Threshold

The statistical variation of γ_d is introduced by the two clutter terms (k_2 and k_3) since the target k_1 is deterministic in the case of point targets. As explained in the second chapter, the amplitude square of the components (complex Gaussians) are Exponential random variables (Oliver and Quegan 1998). Additionally, the average of exponential distributions with equal mean reduces the variance of the resulting random variable. In particular, the sum of Exponentials is a Gamma (Γ)

Fig. 4.4 Deterministic
detector (for different values
of $|b|/|a|$)

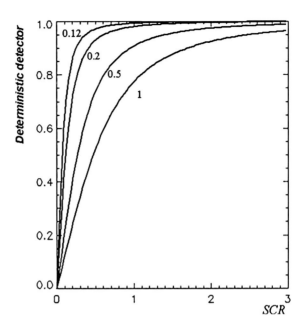

distribution (Gray and Davisson 2004; Papoulis 1965). In the case of independent
and identically distributed (*iid*) samples, the variance of the resulting Γ is divided
by the number of samples considered in the average. Therefore, if an infinite
number of homogeneous and independent samples were available, the variance
would be zero. This suggests that, in order to achieve a deterministic detector, we
could substitute the finite average operator $\langle . \rangle$ with the expected value $E[.]$
(Papoulis 1965). In the following, the resulting mathematical expression will be
regarded as *deterministic* detector. Considering the detector works with high
values of coherence, the latter assumption (i.e. $\langle . \rangle \to E[.]$) is relatively easily
fulfilled even for a 5×5 window size (as it will be shown in the following) (Touzi
et al. 1999). The expression of the deterministic detector is

$$\gamma_{\text{det}} = \frac{1}{\sqrt{1 + \dfrac{|b|^2}{|a|^2} \dfrac{1}{SCR_{d2}} + \dfrac{|c|^2}{|a|^2} \dfrac{1}{SCR_{d3}}}} \tag{4.53}$$

$$SCR_{d2} = E\left[|k_1|^2\right] \Big/ E\left[|k_2|^2\right] \quad \text{and} \quad SCR_{d3} = E\left[|k_1|^2\right] \Big/ E\left[|k_3|^2\right] \tag{4.54}$$

After the perturbed target $\underline{\omega}_P$ (i.e. a, b and c) is fixed, Eq. 4.53 is an expression
related solely to the asymptotic signal to clutter ratios (SCR_d).

Figure 4.4 represents the plot of the deterministic detector, where the value of
the *RedR* is varied. Please note that the mean curve in Fig. 4.2 overlaps almost
perfectly with the one in Fig. 4.4 (for $|b|/|a| = 0.5$), confirming that the detector
assembled with the power terms is not biased.

To be precise the two curves do not overlap perfectly, since in Fig. 4.2 the average curve is calculated as $\langle \gamma_d(SCR) \rangle$. Assuming that the number of samples is big enough, we could say $\langle \gamma_d \rangle \approx E[\gamma_d]$. On the other hand, in Fig. 4.4, $\gamma_d(SCR_d) = \gamma_d(E[SCR])$ is calculated. In general,

$$f(E[x]) \neq E[f(x)], \tag{4.55}$$

where x is a general random variable (Papoulis 1965; Krantz 1999).

However, in our special case the function is monotone concave and the Jensen inequality can be applied:

$$f(E[x]) \geq E[f(x)] \tag{4.56}$$

The equality is fulfilled when the function f is linear or the distribution x is degenerate (e.g. impulsive distribution). The linearity of a function can be related with its curvature. Moreover, we can restrict the linearity property locally based on the spreading of the random variable x (the smaller the variance, the smaller is the local interval).

As is clear from Fig. 4.4, after the saturation, the curvature is almost zero and the function can be easily approximated as linear. This is not true for the middle part of the plot, where the curvature is more consistent. Besides, the second hypothesis about degenerate distribution (i.e. with zero variance) can be applied. After the averaging process the Γ distribution obtained has a relatively small variance, making the curve look more linear. In order to test this property we simulated two detectors with $RedR = 0.25$ and two window sizes of 5×5 and 9×9. The difference between the averaged detector and the deterministic one for a $SCR = 2$ is 0.001 for 5×5 and 0.0005 for 9×9. This demonstrates that the deterministic detector represents a useful tool to analyse the parameters.

In a first attempt, the threshold can be selected using the deterministic detector on the base of the SCR to be detected. This is not the optimal solution (as explained in the next chapter) and it is only intended to provide a general idea about the threshold selection. Figure 4.4 allows some consideration of the $RedR$ as well. The detector increases when the ratio is reduced (the clutter terms are lower). Regarding the choice of the ratio, a small value reduces the variance (since we work with higher values of modified coherence), however the range of discrimination between targets is reduced (the curve flattens earlier).

Regarding the choice of the SCR to detect, the dispersion equation obtained in Appendix 2 can be used. If $\underline{k} = \left[(\sigma + \Delta\sigma') e^{j(\varphi_1' + \Delta\varphi_1')}, \ \Delta\sigma_2' e^{j\varphi_2'}, \ \Delta\sigma_3' e^{j\varphi_3'} \right]^T$ is a normalised scattering vector (the normalisation makes the polarimetric information more apparent) after the change of basis, the dispersion equation can be written as:

$$\frac{\left\langle (\Delta\sigma_2')^2 \right\rangle + \left\langle (\Delta\sigma_3')^2 \right\rangle}{\left\langle (\sigma + \Delta\sigma')^2 \right\rangle} = \frac{1}{SCR} \leq \frac{1}{RedR} \left(\frac{1}{T^2} - 1 \right) \tag{4.57}$$

This expression combines the effect of threshold T and $RedR$ illustrating the collection of targets which will be detected by the algorithm. Equation 4.57 can be used to set the SCR of interest.

Once the $RedR$ is fixed the threshold can be set. For very dominant targets, the detection is easier, hence the minimisation of false alarm is the central point. Therefore, a higher SCR can be chosen (this leads to a higher threshold). On the other hand, if embedded (e.g. foliage penetration FOLPEN) (Fleischman et al. 1996) or weak targets (with low total backscattering) are to be detected, a lower SCR must be selected, consequently a lower threshold must be applied (Kay 1998; Li and Zelnio 1996). The effect of the threshold selection will be clearly visible in the validation chapter.

Related with the detection of weak targets, a relevant property of the algorithm is that the detectability is not directly dependent on the total power scattered by the target (span of the scattering matrix or trace of the covariance matrix), but exclusively on the reciprocal weight of the scattering components. In order to prove this property, we can multiply the matrix $[P]$ by a real positive scalar C. The resulting detector will not change:

$$\gamma_d = \frac{\left|\underline{\omega}_T^{*T} C[P]\underline{\omega}_P\right|}{\sqrt{(\underline{\omega}_T^{*T} C[P]\underline{\omega}_T)(\underline{\omega}_P^{*T} C[P]\underline{\omega}_P)}} = \frac{C}{C} \frac{\left|\underline{\omega}_T^{*T}[P]\underline{\omega}_P\right|}{\sqrt{(\underline{\omega}_T^{*T}[P]\underline{\omega}_T)(\underline{\omega}_P^{*T}[P]\underline{\omega}_P)}}. \quad (4.58)$$

The threshold reduction for weak targets is a consequence of the noise effect, which disturbs the polarimetric characteristics. In order to prove this property, a simulation was performed with absence of clutter and just additive white uncorrelated noise considered as complex Gaussian zero mean (Kay 1998):

$$\underline{k}' = \underline{k}_1 + \underline{n},$$

$$k_1' = k_1 + n_1, \quad k_2' = n_2 \quad \text{and} \quad k_3' = n_3. \quad (4.59)$$

The Signal to Noise Ratio (SNR_γ) can be calculated as

$$SNR_\gamma = \frac{\left\langle |k_1 + n_1|^2 \right\rangle}{\left\langle |n_2|^2 \right\rangle + \left\langle |n_3|^2 \right\rangle}. \quad (4.60)$$

where the first component of the scattering vector is interpreted by the detector as target of interest, even if it contains a noise component. As a result, a threshold of 0.98 is required in order to detect a target embedded in white noise with SNR_γ of about 1 dB and of 0.88 for -10 dB SNR_γ. The positive performance in detecting weak targets is a consequence of the spreading of polarimetrically white noise over all the components. Therefore, part of the noise power (statistically a third) contributes to the target of interest, making the system more robust. In order to check this property we can compare the SNR_γ with the classical definition of the Signal to Noise Ratio, SNR.

$$SNR = \frac{\left\langle |k_1|^2 \right\rangle}{\left\langle |n_1|^2 \right\rangle + \left\langle |n_2|^2 \right\rangle + \left\langle |n_3|^2 \right\rangle} \leq \frac{\left\langle |k_1 + n_1|^2 \right\rangle}{\left\langle |n_2|^2 \right\rangle + \left\langle |n_3|^2 \right\rangle} = SNR_\gamma.$$

(4.61)

In other words, the apparent *SNR* is higher. We will come back to the concept of white clutter in the next chapter.

4.5.2 Perturbed Target Selection: RedR

In the previous formulation a tacit hypothesis has been employed: $|b| = |c|$. The aim of this section is to evaluate the effects of $|b| \neq |c|$. The components of ω_P are not independent, since $|a|^2 + |b|^2 + |c|^2 = 1$ (ω_P is a normalised vector). In order to manifest the role played by b and c, an example is provided. If the deterministic scattering vector, over the entire averaging window, is $\underline{k} = \kappa[a', b_0, 0]$, the power of the components can be calculated as

$$\left\langle |k_1|^2 \right\rangle = \left\langle |\kappa a'|^2 \right\rangle = |\kappa a'|^2 = |\kappa|^2 |a'|^2,$$

(4.62)

$$\left\langle |k_2|^2 \right\rangle = \left\langle |\kappa b_0|^2 \right\rangle = |\kappa b_0|^2 = |\kappa|^2 |b_0|^2,$$

$$\left\langle |k_3|^2 \right\rangle = 0,$$

where

$$|b_0| = \sqrt{1 - |a'|^2}.$$

If the scattering mechanisms for target and perturbed target are chosen as

$$\underline{\omega}_P = [a, 0, c_0], \quad \underline{\omega}_T = [1, 0, 0],$$

(4.63)

where

$$|c_0| = \sqrt{1 - |a|^2},$$

the detector will be

$$\gamma_d(\underline{\omega}_T, \underline{\omega}_P) = \frac{1}{\sqrt{1 + \dfrac{0}{|a|^2} \dfrac{|b_0|^2}{|a'|^2} + \dfrac{|c_0|^2}{|a|^2} \dfrac{0}{|a'|^2}}} = 1.$$

(4.64)

In conclusion, the orthogonality (or in general the geometrical relationship) between the clutter components of $\underline{k}$ and $\underline{\omega}_P$ can bias the detector. In particular, the projection of $\underline{k}$ and $\underline{\omega}_P$ on the complex plane of the clutter (plane orthogonal to $\underline{\omega}_T$) can be represented by

$$[P_c]\underline{k} = [\underline{0},\ \underline{\omega}_{C2},\ \underline{\omega}_{C3}]\underline{k} = \begin{bmatrix} 0 & 0 & 0 \\ 0 & 1 & 0 \\ 0 & 0 & 1 \end{bmatrix}\underline{k} = \underline{k}^c = k_2\underline{\omega}_{C2} + k_3\underline{\omega}_{C3}, \quad (4.65)$$

$$[P_c]\underline{\omega}_P = [\underline{0},\ \underline{\omega}_{C2},\ \underline{\omega}_{C3}]\underline{\omega}_P = \begin{bmatrix} 0 & 0 & 0 \\ 0 & 1 & 0 \\ 0 & 0 & 1 \end{bmatrix}\underline{\omega}_P = \underline{\omega}_P^c = b\underline{\omega}_{C2} + c\underline{\omega}_{C3},$$

$$(4.66)$$

where $[P_C]$ represents the projection matrix on the complex plane of the clutter (Rose 2002; Strang 1988). The geometrical relationship between $\underline{k}^c$ and $\underline{\omega}_P^c$ influences the final coherence since the second part of the denominator can be seen as the inner product of these two vectors restricted to the positive quadrant (since all the quantities appear as squared amplitude):

$$\gamma_d = \frac{1}{\sqrt{1 + \dfrac{\left(\underline{C}^k\right)^{*T}\underline{C}^\omega}{|a|^2\left\langle|k_2|^2\right\rangle}}} = 1 \quad (4.67)$$

where

$$\underline{C}^k = \left\langle|k_2|^2\right\rangle\underline{\omega}_{C2} + \left\langle|k_3|^2\right\rangle\underline{\omega}_{C3} \quad \text{and} \quad \underline{C}^\omega = |b|^2\underline{\omega}_{C2} + |c|^2\underline{\omega}_{C3}.$$

As a result, when they are orthogonal the inner product is zero and the coherence is one independently of the value of the target.

A relationship between b and c is investigated which makes the detector not biased. It can be demonstrated that this choice is $|b| = |c|$. Since $\underline{C}^k$ and $\underline{C}^\omega$ are positive vectors the only way to be orthogonal is that they represent the two positive axes of the clutter plane. Moreover, the only choice which gives a fair weighting between the two components is the one when the two components of $\underline{\omega}_P^c$ are equal. Please note, when we have a priori hypothesis about the target to detect we could be interested in lowering one component more than the other. For instance, when one clutter component is more likely to accompany our target (while the second component is always low), we could decide to have $|b| \neq |c|$. However, in this thesis the more general case of absence of a priori hypothesis will be considered.

Adopting $|b| = |c|$, the detection is unbiased. In order to prove it, we consider a general deterministic target as $\underline{k} = \kappa[a',\ b',\ c']$ (same value over the entire averaging window). After algebraic manipulations we have:

$$\gamma_d = \frac{1}{\sqrt{1 + \frac{|b|^2}{|a|^2|a'|^2}\left(|b'|^2 + |c'|^2\right)}}. \tag{4.68}$$

Equation 4.68 states that the total (normalized) power of the clutter components is contained in $|b'|^2 + |c'|^2$, it does not matter which is stronger between b' and c', and the bias is removed.

Concerning the physical feasibility of this operation: it is always possible to match $|b| = |c|$ with a rotation of the perturbed target around the axis representing the target of interest (i.e. a rotation in the clutter complex plane). The transformation does not require a change of phase since we are interested in amplitudes and the phases of b and c can be arbitrary (Cloude 1995a, 2009).

Mathematically, if $\underline{\widehat{\omega}}_P$ is the scattering mechanisms obtained by perturbation of $\underline{\omega}_T$, the rotation can be performed with the unitary matrix (Hamilton 1989)

$$\begin{bmatrix} 1 & 0 & 0 \\ 0 & \cos(\varphi) & -\sin(\varphi) \\ 0 & \sin(\varphi) & \cos(\varphi) \end{bmatrix} \begin{bmatrix} a' \\ b' \\ c' \end{bmatrix} = \begin{bmatrix} a \\ b \\ c \end{bmatrix} = \underline{\omega}_P, \tag{4.69}$$

where the a component is not modified by the rotation (hence $a' = a$).

4.6 Algorithm Implementation

In the previous sections the mathematical formulation of the detector has been carried out, resulting in a final mathematical expression. However, its practical implementation was left out and will be the topic of the current section. As will be detailed in the following, the final algorithm is fast (low time consuming) since it is based on a relatively small number of multiplications (hence it could be implemented in a real time scenario).

4.6.1 Gram–Schmidt Ortho-normalisation

The final expression of the detector in Eq. 4.37 is dependent on the basis used to represent the vectors $\underline{\omega}_T$ and $\underline{\omega}_P$. Specifically, a change of basis making $\underline{\omega}_T = [1, 0, 0]^T$ was exploited. In this basis, the target to detect is present exclusively in the k_1 component (i.e. k_2, k_3 are clutter). Subsequently, the final expression of the detector was simplified in

$$\gamma_d = \frac{1}{\sqrt{1 + \frac{|b|^2}{|a|^2}\frac{P_{C2}}{P_T} + \frac{|c|^2}{|a|^2}\frac{P_{C3}}{P_T}}}. \tag{4.70}$$

In a straightforward implementation the change of basis can be derived solving a system of three complex equations. The operation can be seen as two rigid rotations and a change of phase. The change of phase is just one because of the symmetry of the system, since after the two rotations only one component is different from zero.

In order to make the processing easier, we are looking for an alternative way to find the elements of the detector namely P_T, P_{C2} and P_{C3}. One solution considers a Gram–Schmidt ortho-normalisation (Strang 1988; Hamilton 1989; Rose 2002) which sets $\underline{\omega}_T$ as one axis of the new basis of the target space. This new basis will be composed by three unitary vectors $\underline{u}_1 = \underline{\omega}_T$, $\underline{u}_2 = \underline{\omega}_{C2}$ and $\underline{u}_3 = \underline{\omega}_{C3}$, where again $\underline{\omega}_{C2}$ and $\underline{\omega}_{C3}$ are orthogonal to $\underline{\omega}_T$ and they lie on the clutter complex plane. Subsequently, P_T, P_{C2} and P_{C3} can be calculated with the averaged squared amplitude of the inner product of the observable k on the three vectors of the basis:

$$P_T = \left\langle \left| \underline{k}^T \underline{u}_1 \right|^2 \right\rangle, \quad P_{C2} = \left\langle \left| \underline{k}^T \underline{u}_2 \right|^2 \right\rangle \quad \text{and} \quad P_{C3} = \left\langle \left| \underline{k}^T \underline{u}_3 \right|^2 \right\rangle \quad (4.71)$$

With this procedure, the process that makes the detector a pure mathematical operator is completed.

Looking at Eq. 4.71, the feasibility and uniqueness of the single target detection is apparent. The power terms are obtained with the projection of the scattering vector on the scattering mechanism. When all the five parameters of the scattering vector are employed, the single target can be completely characterised (Cloude 1992, 1995). Moreover the averaging operator allows us to take into account the partial nature of the clutter. Details on the uniqueness of the detection are presented in Appendix 2.

4.6.2 Flow Chart

Figure 4.5 shows the logical flow chart, in a straightforward attempt to implement the detector. The first step is the definition of the perturbed target starting from the target to detect. Subsequently, target, perturbed target and scattering vector (i.e. observable) are represented in the basis that makes $\underline{\omega}_T = [1, 0, 0]^T$. The perturbation analysis can be performed with several methodologies. Afterwards, the weighting matrix $[A]$ or equivalently the metric matrix $[P]$ is formed starting from the scattering vector in the new basis. Then, the weighted inner product between target and perturbed target is estimated and normalised. The result is the detector γ_d. A threshold on the detector γ_d concludes the algorithm. The final mask will be 0 if the detector is below the threshold or the values of the coherence γ_d if this is above the threshold. Such a mask is preferred to a standard "1–0" mask since the dominance of the target can be appreciated to some extent.

Fig. 4.5 Flow chart used to logically implement the detector algorithm

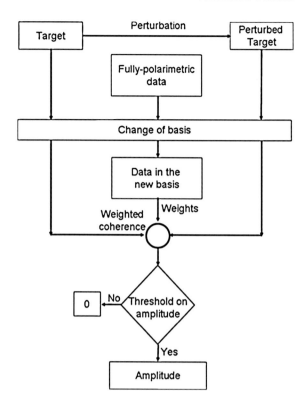

In the previous section, a Gram–Schmidt ortho-normalisation was proposed to solve the problem of finding the change of basis able to make $\underline{\omega}_T = [1,\ 0,\ 0]^T$, simplifying the final detector implementation (Rose 2002; Strang 1988). In Fig. 4.6 the flow chart of an algorithm employing the ortho-normalisation is illustrated. The main divergence is in the first steps. Starting from the expression of the target to detect $\underline{\omega}_T$ in any basis, Gram–Schmidt is applied deriving the two orthogonal clutter components $\underline{\omega}_{C2}$ and $\underline{\omega}_{C3}$. The three power terms are estimated projecting the vector k on the three vectors resulting of the ortho-normalisation.

4.7 Target to Detect

The theoretical formulation asserts that the detector can be focused on any single target as long as its representation is known. Several parameterisations can be used to characterise the target: Polarization Fork, Huynen coherent decomposition, or α model (Cloude 2009; Huynen 1970; Kennaugh and Sloan 1952; Lee and Pottier 2009). However, in order to test the algorithm over real data the

Fig. 4.6 Flow chart used to practically implement the detector algorithm

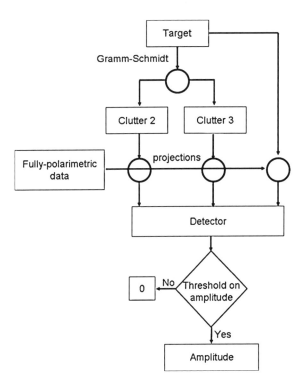

detection must be specialised and aimed at specific targets. Once the scattering mechanism for the target to detect is found in the basis employed by the parameterisation, it must be converted in the detector basis. As explained before an easier way considers the Gram–Schmidt ortho-normalisation.

4.7.1 Standard Single Targets

By "standard targets" we mean those target typologies widely treated in the literature. Generally, their polarimetric description is relatively simple to extract, besides, they are rather common on a SAR image. For this reason, they will allow a reasonably broad validation in the next chapter. In this section, these polarimetric targets will be presented utilising their polarisation fork. A large number of examples of real physical targets related to standard targets will be presented in the validation (Chaps. 6 and 7).

As a first attempt multiple reflections (odd and even bounces) and oriented dipoles (horizontally and vertically) will be analysed. Figure 4.7 represents the Poincaré sphere with characteristic polarisations for the targets considered. The x symbolises the Cross-pol Nulls and the circle the Co-pol Nulls or Cross-pol Max.

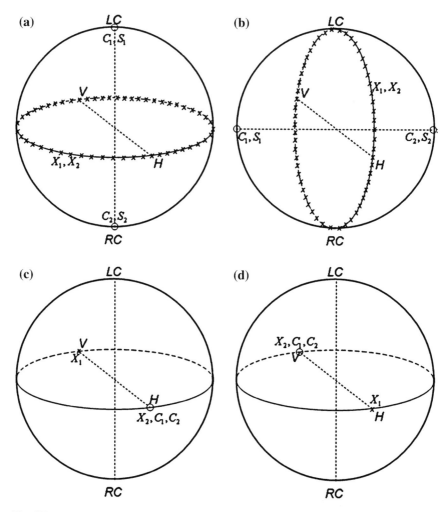

Fig. 4.7 Poincare representation of single targets **a** odd-bounce, **b** even-bounce, **c** vertical dipole, **d** horizontal dipole

As explained in the second chapter, reflecting targets are special targets and this is reflected on their Polarisation Fork as well (Cloude 1987; Huynen 1970; Kennaugh and Sloan 1952).

In particular, the Cross-pol Nulls for odd-bounces are all the linear polarizations (Fig. 4.7a). This means that there is an infinite number of Cross-pol Nulls while general targets have just two of them. Any linear polarisation incident on an odd-bounce (which can be a surface or a sphere) will be reflected with no depolarisation. The Co-pol Nulls or Cross-pol Max are the circular polarizations. The equality between Co-pol Nulls or Cross-pol Max is a characteristic of the reflections.

In Fig. 4.7b, the Cross-pol Nulls for even bounce are the vertically or horizontally oriented polarizations from linear to circular (on the Poincaré sphere they represent a circle passing through linear H, linear V and circular polarizations). Hence, if a horizontal polarisation (the ellipticity does not matter) is incident on a horizontal corner, it will be reflected without cross scattering. The Co-pol Nulls or Cross-pol Max are the linear polarizations oriented at 45°. When a 45° linear polarisation is transmitted the polarisation returning to the sensor will be completely orthogonal, due to the change of sign of the horizontal component of the field.

The horizontal (vertical) dipole has only one Cross-pol Null different from zero, the horizontal (vertical) linear polarisation, and the Co-pol Nulls and second Cross-pol Null (which has amplitude zero) is the vertical (horizontal) linear polarisation (Figs. 4.7c, d). Dipoles are degenerate eigenvectors, and one eigenvalue is zero.

References

Bebbington DH (1992) Target vectors: spinorial concepts. In: Proceedings 2nd international workshop on radar polarimetry, IRESTE, Nantes, France, pp 26–36

Boerner WM (2004) Basics of Radar Polarimetry. RTO SET Lecture Series

Boerner WM, El-Arini MB, Chan CY, Mastoris PM (1981) Polarization dependence in electromagnetic inverse problems. IEEE Trans Antenna Propag 29:262–271

Cameron WL, Leung LK (1990) Feature motivated polarization scattering matrix decomposition. Record of the IEEE international radar conference pp 549–557

Cherniakov M (2008) Bistatic radar: emerging technology. Wiley, Chichester

Cloude SR (1986) Group theory and polarization algebra. OPTIK 75:26–36

Cloude SR (1987) Polarimetry: the characterisation of polarisation effects in EM scatterinig. Electronics engineering department York, University of York, York

Cloude RS (1992) Uniqueness of target decomposition theorems in radar polarimetry. In: Direct and inverse methods in radar polarimetry, pp 267–296

Cloude SR (1995a) Lie groups in EM wave propagation and scattering. In: Baum C, Kritikos HN (eds) Electromagnetic symmetry. Taylor and Francis, Washington, pp 91–142. ISBN 1-56032-321-3

Cloude SR (1995b) Symmetry, zero correlations and target decomposition theorems. In: Proceedings of 3rd international workshop on radar polarimetry (JIPR '95). IRESTE pp 58–68

Cloude SR (2009) Polarisation: applications in remote sensing. Oxford University Press, Oxford. 978-0-19-956973-1

Cloude SR, Pottier E (1996) A review of target decomposition theorems in radar polarimetry. IEEE Trans Geosci Remote Sens 34:498–518

Cloude SR, Pottier E (1997) An entropy based classification scheme for land applications of polarimetric SAR. IEEE Trans Geosci Remote Sens 35:68–78

Fleischman JG, Ayasli S, Adams EM (1996) Foliage attenuation and backscatter analysis of SAR imagery. IEEE Trans Aerosp Electron Syst 32:135–144

Franceschetti G, Lanari R (1999) Synthetic aperture radar processing. CRC Press, Boca Raton

Freeman A (1992) SAR calibration: an overview. IEEE Trans Geosci Remote Sens 30:1107–1122

Germond A-L, Pottier E, Saillard J (2000) Ultra-wideband radar technology: bistitic radar polarimetry theory. CRC

Gray RM, Davisson LD (2004) An introduction on statistical signal processing. Cambridge University Press, Cambridge

Hamilton AG (1989) Linear algebra: an introduction with concurrent examples. Cambridge University Press, Cambridge

Hippenstiel RD (2002) Detection theory: applications and signal processing. CRC Press, Boca Raton

Huynen JR (1970) Phenomenological theory of radar targets. Delft, Technical University, Delft

Kay SM (1998) Fundamentals of statistical signal processing. vol 2, Detection theory. Prentice Hall, Upper Saddle River

Kennaugh EM, Sloan RW (1952) Effects of type of polarization on echo characteristics. Ohio state University, Research Foundation Columbus, Quarterly progress reports (In lab)

Krantz SG (1999) Jensen's inequality handbook of complex variables. Birkhäuser, Boston

Krogager E, Czyz ZH (1995) Properties of the sphere, di-plane and helix decomposition. In: Proceedings of 3rd international workshop on radar polarimetry, IRESTE. University of Nantes, France, pp 106–114

Lee JS (1986) Speckle suppression and analysis for synthetic aperture radar images. SPIE Opt Eng 25:636–643

Lee JS, Pottier E (2009) Polarimetric radar imaging: from basics to applications. CRC press/ Taylor Francis Group, Boca Raton

Li J, Zelnio EG (1996) Target detection with synthetic aperture radar. IEEE Trans Aerosp Electron Syst 32:613–627

López-Martínez C, Fàbregas X (2003) Polarimetric SAR speckle noise model. IEEE Trans Geosci Remote Sens 41:2232–2242

Marino A, Woodhouse IH (2009) Selectable target detector using the polarization fork. In: IEEE Int Geos and RS Symp IGARSS 2009

Marino A, Cloude S, Woodhouse IH (2009) Polarimetric target detector by the use of the polarisation fork. In: Proceedings of 4th ESA international workshop, POLInSAR 2009

Marino A, Cloude SR, Woodhouse IH (2010) A polarimetric target detector using the huynen fork. IEEE Trans Geosci Remote Sens 48:2357–2366

Mott H (2007) Remote Sensing with Polarimetric Radar. Wiley, Hoboken, New Jersey

Oliver C, Quegan S (1998) Understanding synthetic aperture radar images. Sci Tech Publishing, Inc., Raleigh

Papoulis A (1965) Probability random variables and stochastic processes. McGraw Hill, New York

Rose HE (2002) Linear algebra: a pure mathematical approach. Birkhauser, Berlin

Strang G (1988) Linear algebra and its applications. 3rd edn. Thomson Learning, Farmington Hills

Touzi R, Lopes A, Bruniquel J, Vachon PW (1999) Coherence estimation for SAR imagery. IEEE Trans Geosci Remote Sens 37:135–149

Willis NJ (2005) Bistatic radar. SciTech, Releigh

Chapter 5
Polarimetric Detector Statistics

5.1 Introduction

In the previous section a physical/geometrical approach was executed to develop the polarimetric detector. In order to set preliminary detector parameters a deterministic formulation was derived substituting finite with infinite averaging. Unfortunately, the previous procedure is unable to provide information about the detector's statistical performance since the variability is removed completely. The aim of this chapter is to study the detector as a stochastic process and examine its statistical performance (Kay 1998). The complete statistical characterisation of a random variable can be accomplished by calculating its probability density function (*pdf*) analytically. Once the detector can be described as a stochastic process it can be easily compared with other detectors for theoretical validation (Kay 1998; Li and Zelnio 1996; Chaney et al. 1990).

5.2 Analytic Detector Probability Density Function

Any stochastic process can be completely characterised only when information about the random variables which generate it are available. In other words, a priori hypotheses on target and clutter are required to extract the *pdf* (Papoulis 1965; Gray and Davisson 2004). Please note, the algorithm does not need statistical a priori information to perform the detection, since the procedure follows a physical rationale. On the other hand, they are necessary to extract the *pdf* which will be used to optimise the parameter setting. In the following formulation, the term *coloured* defines a clutter showing some polarimetric dependence, while the term *white* designates a clutter independent of the polarisation, hence completely depolarised (i.e. the same power scattered in any polarisation).

A. Marino, *A New Target Detector Based on Geometrical Perturbation Filters for Polarimetric Synthetic Aperture Radar (POL-SAR)*, Springer Theses, DOI: 10.1007/978-3-642-27163-2_5, © Springer-Verlag Berlin Heidelberg 2012

5.2.1 Coloured Clutter Hypothesis

If $\underline{k} = [k_1, k_2, k_3]^T$ is the scattering vector after the change of basis which makes $\underline{\omega}_T = [1, 0, 0]^T$ the k_1 component will represent the target to detect and k_2 and k_3 are the clutter components. In this hypothesis, the target to detect is deterministic (e.g. a point target) and the two clutter components are random variables, specifically Gaussians zero mean (Franceschetti and Lanari 1999; López-Martínez and Fàbregas 2003; Oliver and Quegan 1998):

$$\text{Re}(k_2), \ \text{Im}(k_2), \ \text{Re}(k_3), \ \text{Im}(k_3) \sim N(0, \sigma_g). \tag{5.1}$$

Due to the statistical variability of the generators (i.e. clutter components) the detector γ_d becomes a random variable defined between zero and one (Touzi et al. 1999).

This is the hypothesis assumed in the last chapter since it provides the best picture of the detector where the full power of the clutter contributes in lowering the detector value. As will be explained in the following, this is also the worst case scenario.

The *pdf* of the detector could be derived with a transformation of the generator distributions in the final detector (as presented in Eq. 4.37) (Papoulis 1965). Specifically, the detector is a function of four Gaussian random variables $\gamma_d(\text{Re}(k_2), \text{Im}(k_2), \text{Re}(k_3), \text{Im}(k_3))$, therefore the transformation is $4 \rightarrow 1$. Remarkably, the random variables appear in the detector exclusively as averages of powers. If the clutter is assumed homogeneous with independent realisations (*iid:* independent and identically distributed), power terms are two rescaled Γ distributions:

$$\left\langle |k_2|^2 \right\rangle = P_{C2} \sim \Gamma\left(\frac{\sigma}{N}, N\right) \ \text{and} \ \left\langle |k_3|^2 \right\rangle = P_{C3} \sim \Gamma\left(\frac{\sigma}{N}, N\right), \tag{5.2}$$

where σ is the mean of the single exponential variable (single pixel intensity) and N is the number of samples considered in the averaging window. Furthermore, P_{C2} and P_{C3} are independent of each other (Papoulis 1965; Oliver and Quegan 1998).

The transformation is now simplified into a $2 \rightarrow 1$:

$$\gamma = \frac{1}{\sqrt{1 + RedR_1 \dfrac{P_{C2}}{P_T} + RedR_2 \dfrac{P_{C3}}{P_T}}} = g(P_{C2}, P_{C3}), \tag{5.3}$$

where *RedR* represents the reduction ratio. To derive the *pdf* $f_\Gamma(\gamma)$, the cumulative distribution function (CDF) of γ must be calculated and subsequently differentiated. Considering the complexity of the analytical expression of the clutter terms, we choose not to follow this methodology since the derivation can easily lead to an unsolvable analytical expression.

The problem can be further simplified. If the transformation would be $1 \rightarrow 1$ (one to one) the fundamental theorem of transformation of random variables could be used, reducing drastically the complexity of the calculations (Gray and Davisson 2004).

In the previous chapter, it was shown that the optimum selection for the two *RedR* is $RedR_1 = RedR_2$. Substituting this, the transformation becomes:

$$\gamma = \frac{1}{\sqrt{1 + \dfrac{RedR}{P_T}(P_{C2} + P_{C3})}} = g(P_{C2}, P_{C3}).$$

(5.4)

The two clutter powers can be merged

$$P_C = P_{C2} + P_{C3},$$

(5.5)

resulting in a simplified $1 \rightarrow 1$ transformation

$$\gamma = \frac{1}{\sqrt{1 + RedR\dfrac{P_C}{P_T}}} = g(P_C).$$

(5.6)

The fundamental theorem for transformation of random variable states that given $\gamma = g(P_C)$, the *pdf* of the transformed variable can be calculated as (Papoulis 1965)

$$f_\Gamma(\gamma) = \begin{cases} 0 & \text{the equation } \gamma = g(p_c) \text{ has no solutions} \\ \sum_i \dfrac{f_{P_C}(p_C^i)}{|g'(p_C^i)|} & \text{with } (p_C^i) \text{ solutions of } \gamma = g(p_C), \end{cases}$$

(5.7)

where g' stands for derivative of g

$$g'(P_c) = \frac{dg}{dp_C}.$$

(5.8)

The drawback of this procedure is that it necessitates the calculation of the *pdf* of P_C, which is itself a $2 \rightarrow 1$ transformation. Fortunately, the new random variable P_C can be described rather straightforwardly. The single rescaled Γ distribution represents the average of N independent exponential variables (intensity) as

$$P = \frac{1}{N}\sum_{i=1}^{N} I_i = \frac{1}{N}\sum_{i=1}^{N} |k_i^x|^2,$$

(5.9)

where I stands for pixel intensity and k^x is any of the two clutter components of the scattering vector (i.e. $x = 2, 3$). The *pdf* of P is

$$f_P(P) = \frac{1}{\Gamma(N)}\left(\frac{N}{\sigma}\right)^N p^{n-1} e^{-\frac{Np}{\sigma}},$$

(5.10)

where $\Gamma(N) = (n - 1)!$ is the Γ function and ! is for *factorial*.

In our hypothesis the two variables P_{C2} and P_{C3} are independent and identically distributed (*iid*). Their sum can be written as

$$P_C = P_{C2} + P_{C3} = \frac{1}{N}\left(\sum_{i=1}^{N}|k_i^2|^2 + \sum_{i=1}^{N}|k_i^3|^2\right). \tag{5.11}$$

Considering the variables are *iid* and the intensity of any pixel is independent of all the others, k^2 and k^3 can be substituted with a unique random variable k.

The clutter power becomes:

$$P_C = \frac{1}{N}\sum_{i=1}^{2N}|k_i|^2, \tag{5.12}$$

which is a Γ distribution with different a normalisation factor. Finally, the *pdf* of the clutter components is

$$f_{P_C}(P_C) = \frac{1}{\Gamma(2N)}\left(\frac{N}{\sigma}\right)^{2N}p^{2N-1}e^{-\frac{Np}{\sigma}}. \tag{5.13}$$

Equation 5.13 can be interpreted as the result of a summation property for *iid* Γ distributions (Papoulis 1965):

$$\sum_{j=1}^{L}P_{Cj} \sim \Gamma\left(\sigma, \sum_{j=1}^{L}N_j\right). \tag{5.14}$$

The solution of $\gamma = g(P_C)$ is

$$p_C^1 = \frac{P_T}{RedR}\left(\frac{1}{\gamma^2}-1\right), \tag{5.15}$$

The solution is unique since the function is monotone concave. Considering that γ has no solution external to $[0, 1]$, the *pdf* will be defined only in this interval.

The derivative of $g(p_C)$ is

$$g'(p_C) = -\frac{1}{2}\left(1 + RedR\frac{p_C}{P_T}\right)^{-\frac{3}{2}}\frac{RedR}{P_T}. \tag{5.16}$$

Substituting the solution p_C^1 the derivative becomes

$$g'(p_C^1) = -\frac{1}{2}\frac{RedR}{P_T}\gamma^3. \tag{5.17}$$

Finally, the *pdf* for the detector can be calculated as

$$f_\Gamma(\gamma) = \frac{2}{\Gamma(2N)}\left(\frac{N}{\sigma}\right)^{2N}\frac{1}{\gamma^3}\left(\frac{1}{\gamma^2}-1\right)^{2N-1}\left(\frac{P_T}{RedR}\right)^{2N}e^{-\frac{N}{\sigma}\frac{P_T}{RedR}\left(\frac{1}{\gamma^2}-1\right)}. \tag{5.18}$$

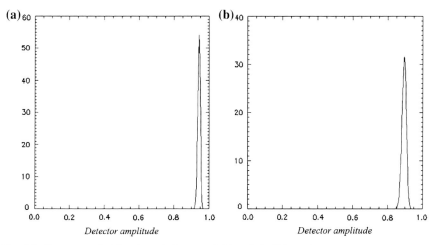

Fig. 5.1 Detector *pdf* with coloured clutter: **a** *SCR* = 2, **b** *SCR* = 1

Table 5.1 Detector parameters	SCR	Window size (N)	RedR	Threshold (T)
	(a) 2	25	0.25	0.95
	(b) 1			

The obtained *pdf* is dependent on the amplitude of the target P_T and clutter 2σ, consequently it is a function of the *SCR*:

$$f_\Gamma(\gamma) = \frac{2}{\Gamma(2N)}\left(\frac{2N}{RedR}\right)^{2N} SCR^{2N}\frac{1}{\gamma^3}\left(\frac{1}{\gamma^2}-1\right)^{2N-1} e^{-\frac{2N}{RedR}SCR\left(\frac{1}{\gamma^2}-1\right)}.$$

$$(5.19)$$

Figure 5.1 shows the *pdf* for two different values of *SCR*. The parameters used to obtain the plots are listed in Table 5.1 (Please note, the threshold parameter will be used in the following).

The *pdf* presents a "bell-like" trend accumulated in a small range of values with consequent modest variability (Papoulis 1965). A similar result was presented in the previous chapter (Fig. 4.2) with the plot of a simulated detector contained in the standard deviation boundaries. The analytical expression of $f_\Gamma(\gamma)$ is in agreement with the value predicted by the mean detector. The particularly small variance makes the punctual values of $f_\Gamma(\gamma)$ higher than 1 (please note the integral is still unitary). Increasing the *SCR*, the *pdf* seems to move right toward 1, reducing its variability (as observed in the previous chapter). In order to test more accurately the dependence on the *SCR*, $f_\Gamma(\gamma)$ can be plotted as function of the *SCR* in a 3 dimensional surface (Fig. 5.2).

The analytical trend follows the one observed in the previous chapter (Fig. 4.2). Increasing the *SCR*, the detector has more probability to be closer to 1 and its

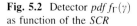

Fig. 5.2 Detector *pdf* $f_\Gamma(\gamma)$ as function of the *SCR*

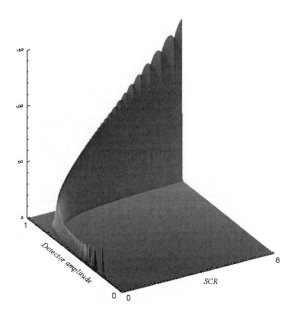

statistical variability decreases, resulting in an increase of the $f_\Gamma(\gamma)$ peak. In the limit (Riley et al. 2006; Mathews and Howell 2006):

$$\lim_{SCR\to\infty} f_\Gamma(\gamma) = \delta(\gamma - 1). \tag{5.20}$$

5.2.2 White Clutter Hypothesis

The coloured clutter hypothesis appears to best characterise the effect of polarimetric clutter, since the entire power is collected in the clutter components of the scattering vector k_2 and k_3. However, this is not the most general hypothesis. The aim of this section is to generalise more the treatment considering clutter equally distributed over all the components of the scattering vector. Such clutter can be associated with generic thermal noise or completely depolarised scattering (e.g. Random Volume composed of spheres) (Fung and Ulaby 1978; Treuhaft and Siqueria 2000; Tsang et al. 1985).

Now, the three random variables which influence the detector are

$$\left\langle |k_{C1}|^2 \right\rangle = P_{C1} \sim \Gamma(\sigma, n),$$

$$\left\langle |k_{C2}|^2 \right\rangle = P_{C2} \sim \Gamma(\sigma, n), \tag{5.21}$$

$$\left\langle |k_{C3}|^2 \right\rangle = P_{C3} \sim \Gamma(\sigma, n),$$

where the scattering vector of the clutter is defined as $\underline{k}_C = [k_{C1}, k_{C2}, k_{C3}]^T$.

Again they are independent and identically distributed. The transformation can be represented as

$$\gamma = \cfrac{1}{\sqrt{1 + RedR_1 \cfrac{P_{C2}}{\langle |k_T + k_{C1}|^2 \rangle} + RedR_2 \cfrac{P_{C3}}{\langle |k_T + k_{C1}|^2 \rangle}}} = g(k_{C1}, P_{C2}, P_{C3}),$$

(5.22)

which is a $3 \rightarrow 1$ transformation. Please note, the power of the target and the first component of clutter cannot be separated as for P_{C2} and P_{C3} since they appear in the same component and they sum coherently. Therefore,

$$\langle |k_T + k_{C1}|^2 \rangle = P_T + P_{C1} + \frac{2}{N} \sum_{i=1}^{N} \mathrm{Re}\{k_T \, k_{iC1}^*\},$$

(5.23)

where N is the window size and k_T is deterministic. $\mathrm{Re}\{k_{C1}\}$ is a Gaussian zero mean, consequently when averaging its variance is divided by N and the term $\frac{2}{N} \sum_{i=1}^{N} \mathrm{Re}\{k_T \, k_{iC1}^*\}$ vanishes for $N \rightarrow \infty$ (since it is zero mean). With the purpose of simplifying the system we assume that the number of realisations N is large enough (e.g. not smaller than 25) to make the cross term negligible in the sum (Riley et al. 2006). Moreover, we assume that the two $RedR$ are equal. After these hypotheses, the resulting function is

$$\gamma = \cfrac{1}{\sqrt{1 + RedR \cfrac{P_{C2} + P_{C3}}{P_T + P_{C1}}}} = \cfrac{1}{\sqrt{1 + RedR \cfrac{P_C}{P_T + P_{C1}}}} = g(P_{C1}, P_C),$$

(5.24)

which is still a $2 \rightarrow 1$ transformation. The derivation with the CDF is not presented since it ends in an expression that is unsolvable analytically. Another approach is pursued. The idea is to introduce another random variable of interest and apply the transformation theorem with a $2 \rightarrow 2$ transformation. This second variable is the power contained in the target component of the scattering vector, defined as $P_{TC} = P_T + P_{C1}$. The latter can be relevant since now the target can be described statistically. A system of equations can be built to contain the two transformations:

$$\Theta = \begin{cases} \gamma = \cfrac{1}{\sqrt{1 + RedR \cfrac{P_C}{P_T + P_{C1}}}} = g(P_{C1}, P_C) \\[2em] P_{TC} = P_T + P_{C1} = h(P_{C1}) \end{cases}.$$

(5.25)

The fundamental theorem for $2 \rightarrow 2$ transformations states:

$$f_{\Gamma P_{TC}}(\gamma, p_{TC}) = \begin{cases} 0 & \text{the system } \Theta \text{ has no solution} \\[1em] \sum_i \cfrac{f_{P_{C1} P_C}(p_{C1}^i, p_C^i)}{|\det(J(p_{C1}^i, p_C^i))|} & \text{with } (p_{C1}^i, p_C^i) \text{ solutions of the system } \Theta \end{cases}$$

(5.26)

where $f_{P_{C1}P_C}\left(p_{C1}^i, p_C^i\right)$ is the joint *pdf* between the two random variables P_{C1} and P_C which are independent (Papoulis 1965). Therefore, the joint *pdf* is factorisable as

$$f_{P_{C1}P_C}\left(p_{C1}^i, p_C^i\right) = f_{P_{C1}}\left(p_{C1}^i\right) f_{P_C}\left(p_C^i\right). \tag{5.27}$$

The matrix $J\left(p_{C1}^i, p_C^i\right)$ is the Jacobian and can be calculated as

$$J(p_{C1}, p_C) = \begin{pmatrix} \dfrac{\partial g}{\partial p_{C1}} & \dfrac{\partial g}{\partial p_C} \\[2ex] \dfrac{\partial h}{\partial p_{C1}} & \dfrac{\partial h}{\partial p_C} \end{pmatrix}. \tag{5.28}$$

If the joint *pdf* is available the individual *pdf* can always be derived by integrating the expression over the entire domain of the other variable.

The solutions of the system Θ are

$$\begin{cases} p_C^1 = \dfrac{P_{TC}}{RedR}\left(\dfrac{1}{\gamma^2} - 1\right) \\[2ex] p_{C1}^1 = P_{TC} - P_T \end{cases}. \tag{5.29}$$

Again the solutions are unique since the two trends are monotonic.

Regarding the Jacobian, two of its four elements can be trivially evaluated

$$\frac{\partial h}{\partial p_{C1}} = 1 \text{ and } \frac{\partial h}{\partial p_C} = 0 \tag{5.30}$$

Consequently the Jacobian becomes

$$J(p_{C1}, p_C) = \begin{pmatrix} \dfrac{\partial g}{\partial p_{C1}} & \dfrac{\partial g}{\partial p_C} \\[2ex] 1 & 0 \end{pmatrix}, \tag{5.31}$$

and the amplitude of the determinant will simply be

$$\left|\det(J(p_{C1}, p_C))\right| = \left|\frac{\partial g}{\partial p_C}\right| = \left|\frac{\partial \gamma}{\partial p_C}\right|. \tag{5.32}$$

The result of the derivative is

$$\frac{\partial \gamma}{\partial p_C} = -\frac{1}{2}\left(1 + RedR\frac{p_C}{P_T + p_{C1}}\right)^{-\frac{3}{2}}\frac{RedR}{P_T + p_{C1}}. \tag{5.33}$$

After the substitution of the solution for the system p_C^1 and p_{C1}^1 we have

$$\frac{\partial \gamma}{\partial p_C}\left(p_C^1, p_{C1}^1\right) = -\frac{1}{2}\frac{RedR}{P_{TC}}\gamma^3. \tag{5.34}$$

which is formally equivalent to the coloured case when the target power P_T is substituted with the actual amount of power in the first component P_{TC}.

The next step is the definition of the *pdf* $f_{P_{C_1}}(p_{C_1}^1)$ and $f_{P_C}(p_C^1)$. f_{P_C} was derived in the previous section. The new variable to characterise is P_{TC}, which can be seen as a $1 \rightarrow 1$ transformation. The derivative is 1 and the *pdf* is merely obtained by substituting the solution in $f_{P_{C_1}}$. Hence,

$$f_{P_{TC}}(p_{TC}) = f_{P_{C_1}}(p_{TC} - P_T), \tag{5.35}$$

$$f_{P_{TC}}(p_{TC}) = \frac{1}{\Gamma(N)} \left(\frac{N}{\sigma}\right)^N (p_{TC} - P_T)^{N-1} e^{-\frac{N}{\sigma}(p_{TC} - P_T)}. \tag{5.36}$$

Equivalently, the transformation could be interpreted as a simple translation of the random variable from 0 to P_T (Riley et al. 2006).

Putting all the results together we can evaluate the joint *pdf*

$$f_{\Gamma P_{TC}}(\gamma, p_{TC}) = \frac{2}{\Gamma(2N)\Gamma(N)} \left(\frac{N}{\sigma}\right)^{3N} \left(\frac{p_{TC}}{RedR}\right)^{2N} (p_{TC} - P_T)^{N-1}$$
$$\cdot \frac{1}{\gamma^3}\left(\frac{1}{\gamma^2} - 1\right)^{2N-1} e^{-\frac{N}{\sigma}\left[\frac{p_{TC}}{RedR}\left(\frac{1}{\gamma^2} - 1\right) + p_{TC} - P_T\right]}. \tag{5.37}$$

In order to extract the individual *pdf* of γ $f_\Gamma(\gamma)$, an integral must be solved

$$f_\Gamma(\gamma) = \int_{P_T}^{\infty} f_{\Gamma P_{TC}}(\gamma, p_{TC}) dp_{TC}. \tag{5.38}$$

Unfortunately, the integral has no analytical solution, however it can be solved numerically (Pearson 1986).

Starting from the analytical expression the characteristic probabilities of the detector can be calculated (in the next section they will be treated more exhaustively). For instance, the probability that the detector is higher than a threshold T given a signal to clutter ratio (*SCR*) can be analytically calculated as

$$P(\gamma \geq T) = \int_T^1 \int_{P_T}^{\infty} f_{\Gamma P_{TC}}(\gamma, p_{TC}) d\gamma dp_{TC}. \tag{5.39}$$

Again a numerical solution must be employed. Therefore, the numerical solution is generally required to calculate the probabilities in any cases.

The *pdf* $f_\Gamma(\gamma)$ after the integration over P_{TC} is shown in Fig. 5.3. The parameters used to define $f_\Gamma(\gamma)$ are the same as Table 5.1. Comparing the *pdf* of coloured and white clutter, it appears that in the case of white clutter the detector is closer to one, given the same *SCR*. In other words, the probability to detect the target is

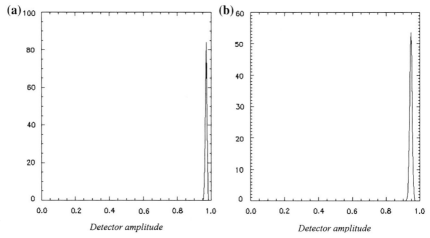

Fig. 5.3 Probability density function of the detector $f_\Gamma(\gamma)$ with white clutter: **a** $SCR = 2$, **b** $SCR = 1$

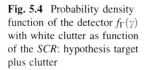

Fig. 5.4 Probability density function of the detector $f_\Gamma(\gamma)$ with white clutter as function of the SCR: hypothesis target plus clutter

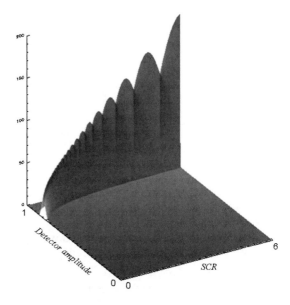

higher since the detector is more likely to pass the threshold. Figure 5.4 depicts the dependence of $f_\Gamma(\gamma)$ on the SCR.

The substantial difference when comparing white against coloured clutter is that in the white case in the absence of a target ($SCR = 0$) the detector is not zero, since there is always power in the target component. Again the *pdf* peak increases with the SCR. A relevant scenario for the calculation of the characteristic probabilities is when the power of the deterministic target is zero (i.e. absence of target). Figure 5.5 presents the plot of $f_\Gamma(\gamma)$ in absence of target.

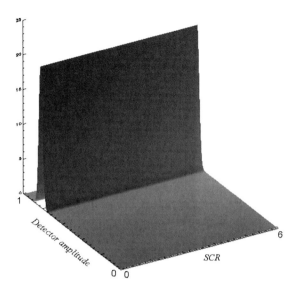

Fig. 5.5 Probability density function of the detector $f_\Gamma(\gamma)$ with white clutter as function of the *SCR*: hypothesis only clutter

5.2.3 General Hypothesis on Clutter

The previous two hypotheses assume that the clutter is distributed only in the clutter components (worst scenario) or equally over all the components. In this section, the most general case will be discussed, where the three clutter components are not identically distributed and the target is still deterministic.

The scattering vector for the clutter can be represented as $\underline{k}_c = [k_{c1}, k_{c2}, k_{c3}]^T$. Clearly, the component $\underline{k}_c = [k_{c1}, 0, 0]^T$ (i.e. without clutter components) cannot be considered clutter since from the physical point of view, it is the target of interest (Cloude 1995). Once again, we would like to stress that this is a statistical evaluation of the performance of the detector, but the definition of target and the selection of the *SCR* must be accomplished with a physical approach (as described in Chap. 4 and Appendix 2).

In this hypothesis there are three power components:

$$P_{Ct} = P_{C1} + P_{C2} + P_{C3} = \left\langle |k_{C1}|^2 \right\rangle + \left\langle |k_{C2}|^2 \right\rangle + \left\langle |k_{C3}|^2 \right\rangle. \qquad (5.40)$$

They are rescaled Γ distributions with parameters:

$$P_{C1} \sim \Gamma\left(\frac{\sigma_1}{n}, n\right), \ P_{C2} \sim \Gamma\left(\frac{\sigma_2}{n}, n\right) \ \text{and} \ P_{C3} \sim \Gamma\left(\frac{\sigma_3}{n}, n\right). \qquad (5.41)$$

where again n is the number of samples in the averaging window and σ_1, σ_2 and σ_3 can be different. Additionally the three components of the clutter are considered independent of each other (please note, now they are not identically distributed).

Again we want to use the fundamental theorem of transformation of random variable and calculate the joint *pdf* (Papoulis 1965). If $P_C = P_{C2} + P_{C3}$ the joint *pdf* of P_{C1} and P_C is

$$f_{P_{C1}P_C}\left(p^i_{C1}, p^i_C\right) = f_{P_{C1}}\left(p^i_{C1}\right) f_{P_C}\left(p^i_C\right). \tag{5.42}$$

$f_{P_{TC}}(p_{TC}) = f_{P_{C1}}(p_{TC} - P_T)$ was calculated previously:

$$f_{P_{TC}}(p_{TC}) = \frac{1}{\Gamma(N)}\left(\frac{N}{\sigma_1}\right)^N (p_{TC} - P_T)^{N-1} e^{-\frac{N}{\sigma_1}(p_{TC} - P_T)}. \tag{5.43}$$

Regarding the distribution of P_C this is the sum of two Γ random variables. Two cases will be considered here:

(a) $\sigma_2 = \sigma_3 = \sigma$

The processing is exactly similar to the white clutter case and P_C has a distribution

$$P_C \sim \Gamma\left(\frac{\sigma}{N}, 2N\right). \tag{5.44}$$

Putting all the results together the expression of the joint *pdf* can be calculated:

$$f_{\Gamma P_{TC}}(\gamma, p_{TC}) = \frac{2}{\Gamma(2N)\Gamma(N)}\left(\frac{N^{3N}}{\sigma_1^N \sigma^{2N}}\right)\left(\frac{p_{TC}}{RedR}\right)^{2N} (p_{TC} - P_T)^{N-1}$$
$$\cdot \frac{1}{\gamma^3}\left(\frac{1}{\gamma^2} - 1\right)^{2N-1} e^{-N\left[\frac{1}{\sigma}\frac{p_{TC}}{RedR}\left(\frac{1}{\gamma^2} - 1\right) + \frac{1}{\sigma_1}(p_{TC} - P_T)\right]}. \tag{5.45}$$

This can be interpreted as the intermediate case between coloured and white clutter. Here, we can separate the random influence on target from the one on the clutter components, giving much more freedom of choice.

This hypothesis is particularly relevant to evaluate the performance of the detector with coloured clutter and thermal noise (as long as the clutter is significantly stronger than the noise). However, this hypothesis is not the most general, since it assumes two statistically similar clutter components.

(b) $\sigma_2 \neq \sigma_3$

This is the most general scenario. In this condition, the summation theorem of Γ distributions is not applicable, since the two distributions are not identically distributed and a transformation $P_C = P_{C2} + P_{C3}$ must be taken into account (i.e. $2 \rightarrow 1$). The CDF of P_C is equal to (Papoulis 1965; Gray and Davisson 2004):

$$F_{P_C}(p_C) = P(P_C \leq p_C) = P(P_{C2} + P_{C3} \leq p_C), \tag{5.46}$$

where the domain where the transformation is formulated is

$$D_{P_C} = \{(p_{C2}, p_{C3}) \in \mathbb{R}^{2+}, p_{C2} + p_{C3} \leq p_C\}. \tag{5.47}$$

The domain can be particularised as:

$$D_{P_C} = \{p_{C2} \in \mathbb{R}^+, p_{C3} \leq p_C - p_{C2}\}. \tag{5.48}$$

By definition the CDF can be seen as the integral of the *pdf* (Papoulis 1965):

$$F_{P_C}(p_C) = \int\int_{D_{P_C}} f_{P_{C2}P_{C3}}(p_{C2}, p_{C3}) dp_{C2} dp_{C3}$$
$$= \int_{-\infty}^{\infty} dp_{C2} \int_{-\infty}^{p_C - p_{C2}} f_{P_{C2}P_{C3}}(p_{C2}, p_{C3}) dp_{C3}. \tag{5.49}$$

To extract the *pdf*, the CDF expression must be differentiated:

$$f_{P_C}(p_C) = \frac{d}{dp_C} \int_{-\infty}^{\infty} dp_{C2} \int_{-\infty}^{p_C - p_{C2}} f_{P_{C2}P_{C3}}(p_{C2}, p_{C3}) dp_{C3}$$
$$= \int_{-\infty}^{\infty} f_{P_{C2}P_{C3}}(p_{C2}, p_C - p_{C2}) dp_{C2}. \tag{5.50}$$

Due to the independence of the two random variables the last expression can be written as:

$$f_{P_C}(p_C) = \int_{-\infty}^{\infty} f_{P_{C2}}(p_{C2}) f_{P_{C3}}(p_C - p_{C2}) dp_{C2}, \tag{5.51}$$

which is a product of convolutions:

$$f_{P_C}(p_c) = f_{P_{C2}}(p_{c2}) * f_{P_{C3}}(p_{c3}), \tag{5.52}$$

where $*$ is the product of convolution (Mathews and Howell 2006; Riley et al. 2006).

An easy way to solve the convolution considers the product of Fourier transformations (Riley et al. 2006):

$$f_{P_C}(p_c) = f_{P_{C2}}(p_{c2}) * f_{P_{C3}}(p_{c3}) = F^{-1}[F[f_{P_{C2}}(p_{c2})]F[f_{P_{C3}}(p_{c3})]]. \tag{5.53}$$

However, the final anti-transformation is unknown.

Clearly the numerical solution is always available (Pearson 1986). The convolution integral can be numerically solved substituting the value p_C^1 and p_{TC}^1 in $f_{P_{C3}}$, since the solutions of the system are known once all the parameters are fixed.

However, we want to attempt a different methodology based on an approximation. The variance of the rescaled $P \sim \Gamma\left(\frac{\sigma}{N}, N\right)$ distributions is

$$VAR[P] = \frac{\sigma^2}{N}, \tag{5.54}$$

hence it is reduced by increasing the number of samples considered (i.e. size of the window). For $N \gg 1$ and small σ the Γ distribution has a trend which can be approximated by a Normal Gaussian distribution with the same variance and mean (Papoulis 1965). This concept follows the central limit theorem and, in general, to have an accurate approximation the variance must be much smaller than the number of samples:

$$\frac{\sigma^2}{N^2} \ll 1 \text{ or } \frac{\sigma}{N} \ll 1. \tag{5.55}$$

In this hypothesis, the distribution of P can be approximated with $P \sim N\left(\sigma, \frac{\sigma^2}{N}\right)$.

If both the clutter components fulfil the approximation we have:

$$P_{C2} \sim N\left(\sigma_2, \frac{\sigma_2^2}{N}\right) \text{ and } P_{C3} \sim N\left(\sigma_3, \frac{\sigma_3^2}{N}\right). \tag{5.56}$$

The approximation is revealed convenient since the distribution of the sum of two Gaussian random variables is still a Gaussian (or in other words, the convolution of two Gaussians is still a Gaussian) (Riley et al. 2006). The resulting variable will have a mean that is equal to the sum of the means and its variance equal to the sum of the variances:

$$P_{C2} + P_{C2} = P_C \sim N\left(\sigma_2 + \sigma_3, \frac{\sigma_2^2 + \sigma_3^2}{N}\right). \tag{5.57}$$

The *pdf* of P_C can be expressed as (Papoulis 1965)

$$f_{P_C}(P_C) = \frac{1}{\sqrt{2\pi\left(\frac{\sigma_2^2 + \sigma_3^2}{N}\right)}} \exp\left[-\frac{(P_C - (\sigma_2 + \sigma_3))^2}{2\left(\frac{\sigma_2^2 + \sigma_3^2}{N}\right)}\right]. \tag{5.58}$$

If this approximation is adopted the joint *pdf* is

$$f_{\Gamma P_{TC}}(\gamma, p_{TC}) = \frac{2}{\Gamma(N)} \frac{p_{TC}}{RedR\gamma^3} \frac{1}{\sigma_1}\left(\frac{N}{\sigma_1}\right)^N (p_{TC} - p_T)^{N-1} \frac{1}{\sqrt{2\pi\left(\frac{\sigma_2^2 + \sigma_3^2}{N}\right)}}$$

$$\cdot \exp\left[-\frac{1}{2\left(\frac{\sigma_2^2 + \sigma_3^2}{N}\right)}\left(\frac{P_{TC}}{RedR}\left(\frac{1}{\gamma^2} - 1\right) - (\sigma_2 + \sigma_3)\right)^2 - \frac{N}{\sigma_1}(P_{TC} - P_T)\right]. \tag{5.59}$$

The obtained expression is a good approximation of the detector when the variance is small with respect to the number of samples. Considering the expression of Γ is evaluated in the solution points of the system, we can adapt the general inequality in Eq. 5.55 to our specific case. When the point of solution is substituted the mean of the distribution is divided by the power in the target component and the *RedR* (even though it is analytically more complicated to isolate the *SCR* parameter in the white clutter hypothesis).

In conclusion, in the point of solution of the system (p_C^1) the accuracy of the approximation depends on the *SCR*:

$$P_C\left(p_C^1\right) \sim \Gamma\left(\frac{\sigma}{N \cdot RedR \cdot P_T}, N\right). \tag{5.60}$$

In order to approximate this distribution with a Gaussian we need

$$\frac{\sigma}{N \cdot RedR \cdot P_T} = \frac{1}{N \cdot RedR \cdot SCR} \ll 1, \tag{5.61}$$

which can be interpreted as

$$SCR \gg \frac{1}{N \cdot RedR}. \tag{5.62}$$

The latter states that the approximation is better when *SCR* and *N* are high. If a 5×5 moving window is used and $RedR = 0.25$, the limit for the *SCR* is around 0.8 which is relatively small as will be shown in the following. Anyway, if we want to use the analytical expression to optimise the algorithm for detection of weaker targets the *RedR* can be adjusted in order to improve the accuracy of the approximation. However, from the physical point of view, targets with *SCR* smaller than 2 are generally not interesting, since they allow an excessive dispersion of the target (Appendix 2).

5.2.4 General Hypothesis on Target and Clutter

In the previous sections, the hypotheses consider the detection of deterministic (point) targets. The physical development of the detector revealed the possibility to detect distributed targets as long as their polarimetric behaviour is single, i.e. the covariance matrix has rank one (this will be validated on real data as well). Distributed targets present speckle variations and they are not deterministic (López-Martínez and Fàbregas 2003; Oliver and Quegan 1998). In this section we want to introduce the expression of the *pdf* when the latter hypothesis is adopted.

The scattering vector now will be $\underline{k} = [k_T + k_{C1}, k_{C2}, k_{C3}]^T$ where:

$$\left\langle |k_T|^2 \right\rangle = P_T \sim \Gamma\left(\frac{\sigma_T}{N}, N\right), \ \left\langle |k_{C1}|^2 \right\rangle = P_{C1} \sim \Gamma\left(\frac{\sigma_1}{N}, N\right), \ \left\langle |k_{C2}|^2 \right\rangle$$
$$= P_{C2} \sim \Gamma\left(\frac{\sigma_2}{N}, N\right) \text{ and } \left\langle |k_{C3}|^2 \right\rangle = P_{C3} \sim \Gamma\left(\frac{\sigma_3}{N}, N\right). \tag{5.63}$$

The system to solve is identical to Θ. The approximation of neglecting the cross terms now is even more effective since k_T is a zero mean random variable as well. However, now the *pdf* of $P_{TC} = P_T + P_{Cl}$ cannot be calculated as the translation of $f_{P_{Cl}}$ and an approximation with the Gaussian has to be performed. Again the accuracy of the approximation is related on the *SCR*. The final expression of the *pdf* is:

$$f_{\Gamma P_{TC}}(\gamma, p_{TC}) = \frac{P_{TC}}{N \pi RedR} \frac{1}{\sqrt{(\sigma_T^2 + \sigma_1^2)(\sigma_2^2 + \sigma_3^2)}} \frac{1}{\gamma^3}$$
$$\cdot \exp\left\{-\frac{n}{2}\left[\frac{\left(P_{TC}\left(\frac{1}{\gamma^2} - 1\right) - \sigma_2 - \sigma_3\right)^2}{RedR\left(\sigma_2^2 + \sigma_3^2\right)} + \frac{\left(P_{TC} - \sigma_T - \sigma_1\right)^2}{\left(\sigma_T^2 + \sigma_1^2\right)}\right]\right\}.$$

$$(5.64)$$

5.3 Probabilities

The $f_\Gamma(\gamma)$ can be used to calculate the probability that the detector is in a defined range of values $\Omega_\Gamma \equiv [\gamma_0, \gamma_1]$ given a fixed $SCR = SCR_0$:

$$P(\gamma_0 \le \gamma \le \gamma_1 | s = SCR_0) = \int_{\gamma_0}^{\gamma_1} f_\Gamma(\gamma | SCR_0) d\gamma, \qquad (5.65)$$

where $s = SCR$.

In the previous equation, the symbol of conditional probability is adopted although the quantity is not precisely a conditional probability (Kay 1998). In fact, the variable s is fixed (i.e. deterministic) and it is not a random variable. However, we decided to adopt the symbol to make the formalism more familiar.

The normalisation property of the *pdf* states that (Papoulis 1965)

$$\int_0^1 f_\Gamma(\gamma | s) d\gamma = 1. \qquad (5.66)$$

Unfortunately, we were not able to extract the analytical expression of the defined integral of $f_\Gamma(\gamma | s)$. Therefore, all the integrals will be performed numerically (Pearson 1986).

The examination of coloured and white clutter hypotheses are provided separately since the probabilities are generally different when different hypotheses are exploited. With the intention of keeping the formulation contained, we did not illustrate the results for the most general hypothesis. However, the trend is supposed to be in between the two extreme cases of coloured and white clutter.

5.3.1 Coloured Clutter Hypothesis

The first step in the calculation of the characteristic probabilities is to define the working hypotheses for the detection (Kay 1998). In this series of tests, we intend to detect the presence or absence of a deterministic target in a statistic clutter spread equally on the clutter components of the scattering vector. In other words, the hypotheses can be summarised in:

$$H_0 : \underline{k} = [k_T, k_{C2}, k_{C3}]^T$$
$$H_1 : \underline{k} = [0, k_{C2}, k_{C3}]^T, \tag{5.67}$$

where k_T represents the target and k_{C2}, k_{C3} are the clutter. The hypotheses can be converted to an expression dependent on the SCR:

$$H_0 : SCR_0 = \frac{\left\langle |k_T|^2 \right\rangle}{\left\langle |k_{C2}|^2 \right\rangle + \left\langle |k_{C3}|^2 \right\rangle} = 2$$

$$H_1 : SCR_1 = \frac{0}{\left\langle |k_{C2}|^2 \right\rangle + \left\langle |k_{C3}|^2 \right\rangle} = 0. \tag{5.68}$$

In particular, when it is not explicitly indicated the SCR of the target is 2. The detection is positive when the detector is above the threshold:

$$\gamma \geq T. \tag{5.69}$$

In detection theory, three probabilities can be identified as particularly relevant in estimating the detector capabilities:

(a) Probability of detection P_D:

This is the probability that the target of interest is present and the detection is positive (Kay 1998). Therefore, the probability P_D can be calculated as

$$P_D(\gamma \geq T | s = SCR_0) = \int_T^1 f_\Gamma(\gamma | SCR_0) d\gamma. \tag{5.70}$$

Clearly, once the threshold is fixed, P_D will be a function of the SCR. This property is tested in Fig. 5.6 where P_D is plotted against the SCR for two different thresholds. Please note, in this thesis the plots of the probabilities depict the SCR in a dB scale, since this makes their interpretation more straightforward. However, in the rest of the thesis the SCR is always shown in its linear scale (unless indicated). Table 5.2 illustrates the selected detector parameters.

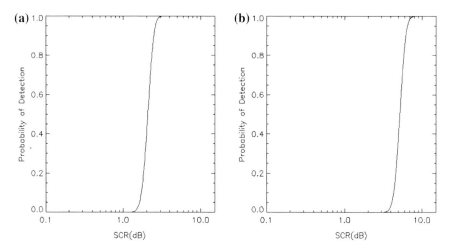

Fig. 5.6 Probability of detection P_D against *SCR* (expressed in dB) for two thresholds. Coloured clutter hypothesis: **a** T = 0.95, **b** T = 0.98

Table 5.2 Detector parameters	SCR	*RedR*	Threshold	Window size
	Variable	0.25	(a) 0.95, (b) 0.98	25

When the threshold increases, the probability P_D is reduced, since a more dominant target is needed to make the coherence pass the threshold (i.e. higher *SCR*).

P_D manifests a peculiar 0–1 (or ON–OFF) trend, where it is approximately 0 till a crossing point where it switches to 1 with a particularly high derivative. This result is favourable for a detector (as will be shown in the following) and it is due to the small variance of $f_\Gamma(\gamma|s)$ (Kay 1998; Papoulis 1965).

The crossing point for $T = 0.95$ is a bit after *SCR* = 2, this means that the mean of the distribution $f_\Gamma(\gamma|s)$ is around *SCR* = 2. The same result was obtained in Fig. 5.1 (or Fig. 4.2).

The second test analyses the reduction ratio (*RedR*) and the window size (Fig. 5.7 and Table 5.3). Increasing the *RedR* the detector becomes more selective and P_D is reduced. Hence, targets must have a higher *SCR* to be detected (the physical explanation of the phenomena is presented in the previous chapter). On the other hand, the effect of decreasing the window size is an increasing variance of the *pdf* (since the *pdf* of random variable generating the detector has higher variance). As a consequence, the probability of detection has a less sharp trend, but the crossing point remains around 0.5, since the mean of $f_\Gamma(\gamma|s)$ is not changed. When the *SCR* (or the coherence value) is lower than the mean, the derivative of P_D increases, while it starts decreasing, when it is higher. In other words, P_D has a flex when *SCR* is equal to the mean.

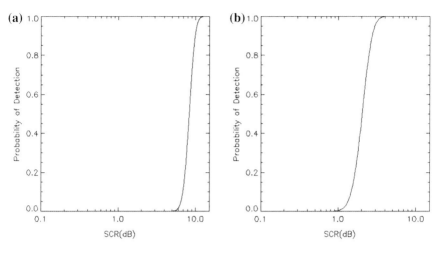

Fig. 5.7 Probability of detection P_D against SCR(dB). Coloured clutter hypothesis. *RedR* and window size are varied (see Table 5.3)

Table 5.3 Detector parameters	SCR	*RedR*	Threshold	Window size
	Variable	(a) 1	0.95	(a) 25
		(b) 0.25		(b) 9

(b) Probability of missed detection P_M:

This is the probability that the target is present (i.e. $s = SCR_0$), but the detector has a negative response $\gamma < T$. It is also called false negative.

$$P_M(\gamma < T | s = SCR_0) = \int_0^T f_\Gamma(\gamma | SCR_0) d\gamma = 1 - P_D. \tag{5.71}$$

As in the previous case, Fig. 5.8 plots P_M as a function of SCR, while the parameters are listed in Table 5.2. Initially, the threshold is varied.

Figure 5.8 clearly displays a property of complementary between P_D and P_M (i.e. $P_M = 1 - P_D$) (Kay 1998). As a consequence both the probabilities must have the same crossing point (e.g. $SCR = 2$ for $T = 0.95$).

Increasing the threshold, P_M increases, since it is more likely to miss targets when they are not sufficiently dominant.

The second experiment (Fig. 5.9 and Table 5.3) studies *RedR* and window size as previously verified for P_D. With a higher *RedR* the filter is more selective and the probability to miss a target is higher. Regarding the window size, the only difference is in the variability of the *pdf* of generator random variables resulting in a less sharp trend of P_M.

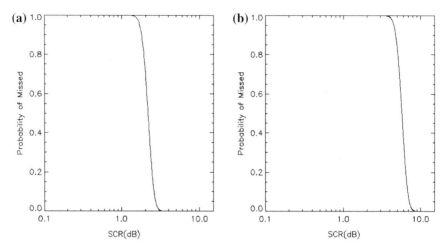

Fig. 5.8 Probability of missed detection P_M against SCR(dB) for two thresholds. Coloured clutter hypothesis: **a** T = 0.95, **b** T = 0.98

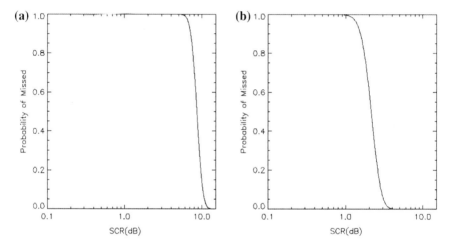

Fig. 5.9 Probability of missed detection P_M against SCR(dB). Coloured clutter hypothesis. *RedR* and window size are varied (see Table 5.3)

(c) Probability of false alarm P_F:

This is the probability that the target of interest is not present $s = 0$, but the detector has a positive response $\gamma \geq T$ (Kay 1998). It is also regarded as false positive.

$$P_F(\gamma \geq T | s = 0). \tag{5.72}$$

The hypothesis of coloured clutter has a strong repercussion on the estimation of this probability. In fact, in absence of target the first component of the scattering vector is completely zero (unless thermal noise is present). Therefore, the detector is deterministically equal to zero,

$$\gamma_d = \frac{1}{\sqrt{1 + RedR\dfrac{P_{C2} + P_{C3}}{0}}} = \frac{1}{\sqrt{1 + \infty}} = 0, \tag{5.73}$$

as well as the probability of false alarm. The plots of P_F are not presented since they are constantly equal to zero.

In this section, the thermal noise is not taken into account for the sake of brevity. However, in order to optimise the detector for a real scenario with coloured clutter, the thermal noise must be added to the treatment. The analytical expression of the *pdf* was presented in Eq. 5.45. and it considers $\sigma_1 \neq \sigma_2 = \sigma_3$. The information about the clutter to noise ratio (CNR) is required

$$CNR = \frac{\sigma_2 + \sigma_3}{\sigma_1}. \tag{5.74}$$

Additionally, in order to be able to use this formulation we need to have $CNR \gg 1$. Moreover, in the case $CNR \ll 1$ the coloured clutter can be neglected and the hypothesis of white clutter can be adopted. For brevity, this treatment is not presented here.

5.3.2 White Clutter Hypothesis

In the case of more general white clutter the working hypotheses are

$$\begin{aligned} H_0 &: \underline{k} = [k_T + k_{C1}, k_{C2}, k_{C3}]^T \\ H_1 &: \underline{k} = [k_{C1}, k_{C2}, k_{C3}]^T, \end{aligned} \tag{5.75}$$

where one of the clutter components is summed to the target.

In terms of *SCR* the hypotheses look similar to the one performed previously:

$$\begin{aligned} H_0 &: SCR_0 = \frac{\left\langle |k_T|^2 \right\rangle}{\left\langle |k_{C2}|^2 \right\rangle + \left\langle |k_{C3}|^2 \right\rangle} = 2 \\ H_1 &: SCR_1 = \frac{0}{\left\langle |k_{C2}|^2 \right\rangle + \left\langle |k_{C3}|^2 \right\rangle} = 0. \end{aligned} \tag{5.76}$$

However, the *SCR* observed by the detector and defined as the ratio of the components is different. The working hypotheses can be translated in an apparent *SCR*:

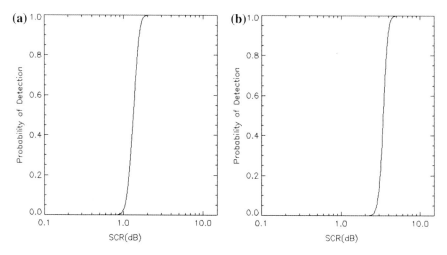

Fig. 5.10 Probability of detection P_D against SCR(dB) for two thresholds. White clutter hypothesis: **a** T $= 0.95$, **b** T $= 0.98$

$$H_0 : SCR_0^A = \frac{\left\langle |k_T|^2 \right\rangle + \left\langle |k_{C1}|^2 \right\rangle}{\left\langle |k_{C2}|^2 \right\rangle + \left\langle |k_{C3}|^2 \right\rangle} \geq 2$$

$$H_1 : SCR_1^A = \frac{\left\langle |k_{C1}|^2 \right\rangle}{\left\langle |k_{C2}|^2 \right\rangle + \left\langle |k_{C3}|^2 \right\rangle} = \frac{1}{2}.$$

(5.77)

(a) Probability of detection:

The probabilities of detection are calculated following the same methodology exploited for coloured clutter. In Fig. 5.10, P_D is plotted against SCR for two different thresholds (the parameters are listed in Table 5.2). Please note, when it is not specified the standard SCR will be considered (as for Eq. 5.76)

Again, an increment of the threshold reduces P_D. In contrast to the coloured counterpart, the crossing point is not at $SCR = 2$, but around smaller values. This is due to the apparent SCR, since part of the clutter power contributes to the detection.

The effect of changing the reduction ratio and window size is shown in Fig. 5.11 (Table 5.3). The effects of these two parameters are equivalent to the coloured hypothesis. Finally, the only substantial difference between coloured and white clutter seems to be the increase in the apparent SCR.

(b) Probability of miss detection:

P_M has the same definition and estimation procedure of the coloured clutter counterpart. Figure 5.12 depicts the results for two different thresholds (Table 5.1).

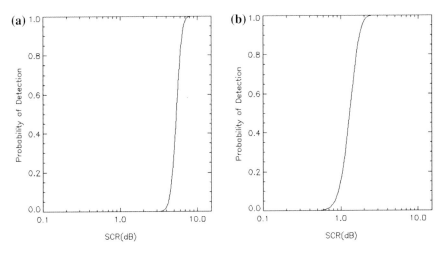

Fig. 5.11 Probability of detection P_D against SCR(dB). White clutter hypothesis. *RedR* and window size are varied (see Table 5.3)

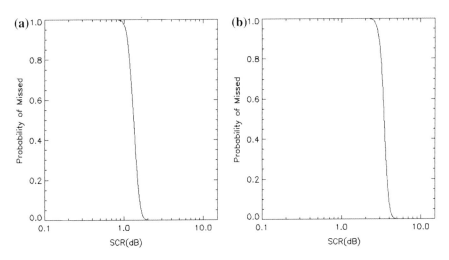

Fig. 5.12 Probability of missed detection P_M against SCR(dB) for two thresholds: White clutter hypothesis: **a** T $= 0.95$, **b** T $= 0.98$

Figure 5.13 shows the plots varying *RedR* and window size (Table 5.3).

All the plots appear to be in agreement with the results of the probability of detection, and in accordance with the coloured case.

(c) Probability of false alarm:

When the clutter is white polarimetrically, the 0 hypothesis (i.e. absence of target) has a correspondent $SCR^A = \dfrac{1}{3}$ and the probability of false alarm is different

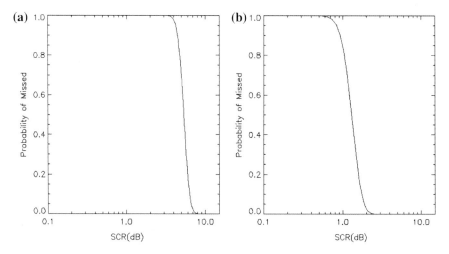

Fig. 5.13 Probability of missed detection P_M against SCR(dB). White clutter hypothesis. $RedR$ and window size are varied (see Table 5.3)

Table 5.4 Probability of false alarm P_F for different parameters

SCR	RedR	Threshold	Window size	P_F
any	0.25	0.95	25	7.2×10^{-15}
any	0.25	0.98	25	1.8×10^{-31}
any	1	0.95	25	4×10^{-39}
any	0.25	0.95	9	8×10^{-6}

from 0. In the absence of a target, SCR^A is constant against the SCR, therefore the P_F will be constant as well. Table 5.4 summarises some examples.

In all the considered scenarios, P_F is particularly small. One relevant issue in designing detectors is to keep P_F small. In this context, the algorithm appears to have promising results (the analysis is provided in the next section).

In the following, we propose some further comments on the results:

(1) If the threshold increases, P_F reduces. It is less likely that some realisations are sufficiently imbalanced to set the detector to the upper the threshold.
(2) Increasing the $RedR$, the coherence becomes lower, reducing P_F and only dominant targets can be detected (Riley et al. 2006):

$$\lim_{RedR \to \infty} \gamma_d = \lim_{RedR \to \infty} \frac{1}{\sqrt{1 + RedR\left(\dfrac{P_{C2} + P_{C3}}{P_T}\right)}} = 0. \qquad (5.78)$$

The effect is similar to pushing the threshold up.

(3) When the number of samples is not sufficient the variability of the clutter amplifies, hence the realisation can be more imbalanced, increasing the P_F. Fortunately, P_F is still low, nevertheless larger windows are generally preferred (Touzi et al. 1999; Lee et al. 1994).

5.4 Receiver Operating Characteristic

Once the analytic statistical expression of the detector are derived and tested, the more fascinating issue of optimising the detector parameters can be tackled. In general the aim of the optimisation is to keep the probability of detection high and the probability of false alarms small. The process commonly involves the selection of the threshold (Kay 1998). The reciprocal weight between P_F and P_D can be visualised for different thresholds providing a direct representation of the detector performances. The latter is regarded as receiver operative characteristic (ROC). The relevance of the ROC is that it can be exploited to compare the statistics of several detectors with different origins (Chaney et al. 1990; Kay 1998; Novak et al. 1993, 1999). The main goal of this section is to generate ROC curves which will be compared with the one presented in Chap. 3.

5.4.1 Coloured Clutter Hypothesis

In the case of coloured clutter, the ROC has no meaning, since the probability of false alarm is always zero. Therefore, the ROC curve appears the same as for a deterministic detector. Clearly the real performances will never be deterministic and the presence of thermal noise will allow the calculation of P_F. As explained previously the *pdf* derived in Eq. 5.45 (i.e. $k_{C1} \neq k_{C2} = k_{C3}$) can be used if the clutter is significantly stronger than the thermal noise.

5.4.2 White Clutter Hypothesis

As presented in the previous section, the false alarm probability is particularly small, however it is not zero and the ROC can be estimated.

Figure 5.14 shows the ROC of the detector for a $SCR = 2$ and an average window of 25 samples (e.g. 5×5). The parameters employed are listed in Table 5.5. In the linear scale, the ROC curve is not visible since it is too close to the point [0, 1]. The latter is obtained punctually by the deterministic detector, since the probability of detection is always 1 and the probability of false alarm is

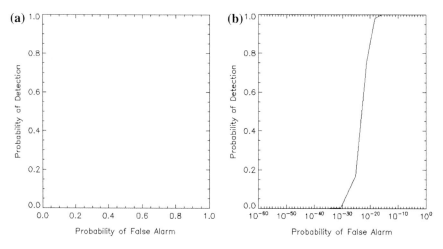

Fig. 5.14 ROC for $SCR = 2$: **a** linear scale, **b** dB scale

Table 5.5 Detector parameters for ROC

SCR	RedR	Threshold	Window size
2	0.25	variable	25

always 0. In order to be able to visualise properly the ROC, we need to display the version where the probability of false alarm is expressed in dB power:

$$P_F^{dB} = 10 \log_{10} P_F. \tag{5.79}$$

In order to obtain P_D sensitively different from 1, P_F must be smaller than 10^{-20}. Considering that commonly requirements of $P_F = 10^{-5}$ are acceptable, the results appear to be superlative (Kay 1998, Chaney et al. 1990).

In actual fact, the performances are so similar to the deterministic detector because the algorithm is not a pure statistical detector. Working with the physics of the scattering the variability can be constrained in two main ways:

(a) It separates target and clutter in the basis set building up a ratio where the variable clutter is normalised to the target.
(b) The averaging reduces the variation of the clutter terms, making them narrower around the mean value (López-Martínez and Fàbregas 2003; Oliver and Quegan, 1998). A larger window results in an even more deterministic detector. On the other hand, a large window degrades the resolution of the system, whereas the resolution plays a central role in target detection. In fact, single targets are generally contained in a small number of pixels and augmenting the dimension of the window the power of the target is spread over a larger area reducing the apparent SCR. In conclusion, a large window is to be preferred in the case of rather extended targets, nevertheless care must be taken for small targets (Novak et al. 1999).

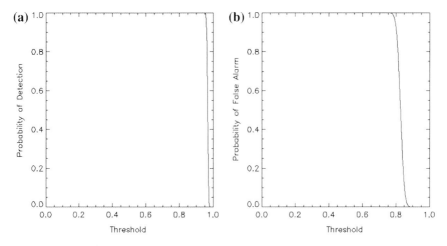

Fig. 5.15 Probability of detection (**a**) and false alarm (**b**) when the threshold is varied

Statistically, the excellent result is due to the sharp variation of the probabilities as function of the threshold (Kay 1998). In Fig. 5.15, P_D, P_F are plotted against the threshold (Table 5.5).

Both the probabilities decrease when the threshold increases (they go from 1 to 0). The explanation of these behaviours are:

(a) P_D: fixed the *SCR*, the probability that the coherence is above the threshold is smaller for higher thresholds (i.e. an higher *SCR* is needed).
(b) P_F: the probability that an unfortunate clutter realisation is higher than the threshold is smaller when the threshold is increased.

Both the trends appear to be extraordinarily sharp with an almost ON–OFF tendency. This creates a region of the plots where P_D is almost one and P_F is almost zero. The optimal threshold can be chosen in this region.

In order to test the detector performances in a more challenging scenario, Fig. 5.16 shows the ROC for a target with $SCR = 1$ (same parameters than Table 5.5). The ROC is still excellent with performances several orders of magnitude better than common requirements. However, compared with the one calculated for $SCR = 2$ it reveals to be lower. Specifically, we have potentially $P_D \approx 1$ with $P_F = 10^{-12}$.

Figure 5.17 tests the dependence of the ROC on the *RedR*. The detector parameters are the same as listed in Table 5.5 except that now $RedR = 1$. Again there is an optimal region for the selection of the threshold and the performances are comparable with the one obtained for $RedR = 0.25$. It seems that the change of *RedR* has no effect on the ROC. A change in ROC seems to be equivalent to a change of the threshold (hence not visible in the graph). Once the threshold is adapted in the new optimal region the performances of the detector remain the same. In order to prove this speculation, Fig. 5.18 illustrates P_D and P_F for $RedR = 1$. The two curves are

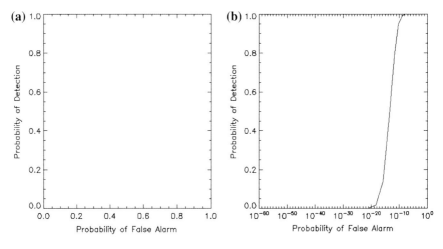

Fig. 5.16 ROC for $SCR = 1$: **a** linear scale, **b** dB scale

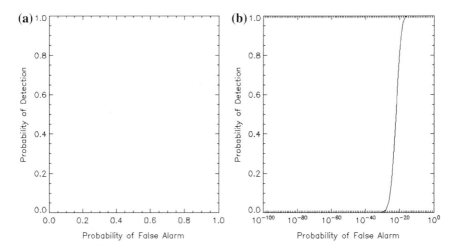

Fig. 5.17 ROC for $RedR = 1$: **a** linear scale; **b** dB scale

shifted leftward, when compared with the case $RedR = 0.25$ and there is still a wide region where the threshold can be selected optimally. In Appendix 2 a similar argument is presented, where $RedR$ and T are linked in a dispersion equation.

The last test concerns the window size (Fig. 5.19). In the previous section it was demonstrated that a decrease of the number of samples amplifies the *pdf* variance. Therefore, we expect it to have effects on the ROC, making the probability trends less sharp. The detection parameters are set as in Table 5.5 except for the window size which is 9 (i.e. 3×3). The ROC moves rightward presenting slightly lower performances, however we can obtain $P_D \approx 1$ with $P_F = 10^{-6}$ which is still one orders of magnitude better than 10^{-5}.

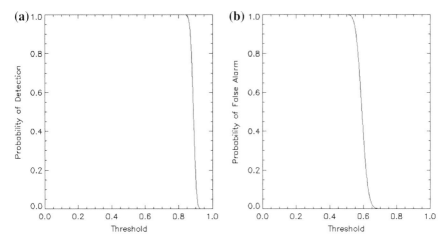

Fig. 5.18 Probability of detection (**a**) and false alarm (**b**) when the threshold is varied

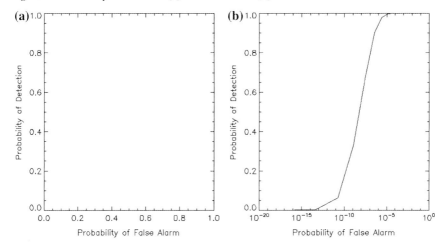

Fig. 5.19 ROC for window size equal to 9: **a** linear scale, **b** dB scale

In the last experiment, the performances of the detector are tested in a particularly challenging scenario (Fig. 5.20). A small window (9 samples) is selected and the examined target has $SCR = 0.5$ (Table 5.6). Please note the limit condition for the detection is $SCR = 0.3$, since below that white clutter is detected. Clearly, in the absence of particular a priori information, this last experiment has more a didactic rather than a practical relevance. In fact, a coherent target which has half the power on the component of interest will be detected as well (Cloude and Pottier 1996). The selection of the SCR must be led by a physical rationale and has to take into account the dispersion equation derived in Appendix 2. In this chapter, the algorithm is examined exclusively from the statistical point of view, with the purpose of optimising its performance. However in the design of the detector, the physical part cannot be neglected.

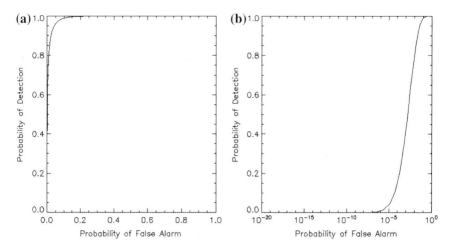

Fig. 5.20 ROC for challenging detection: **a** linear scale, **b** dB scale

Table 5.6 Detector parameters for ROC	SCR	RedR	Threshold	Window size
	0.5	0.25	variable	9

Specifically, converting the *SCR* in an angular distance $\Delta\varphi$ we can define a cone of detected targets (as shown in Appendix 2), we have:

$$SCR = 2 \Rightarrow \Delta\varphi = 35,$$
$$SCR = 1 \Rightarrow \Delta\varphi = 45, \qquad (5.80)$$
$$SCR = 0.5 \Rightarrow \Delta\varphi = 54.$$

An angular distance of 54° is in many cases excessively large. Clearly, it is not to exclude the possibility that some peculiar a priori information could allow the selection of *SCR* = 0.5 without the problem of physical false alarms. The previous calculation of physical dispersion is useful to make some consideration on the hypothesis *SCR* = 2 as well. We judge $\Delta\varphi = 35$ too large for practical detection and in the validation chapter we restrict the angular variation to 15°. On the other hand, in this chapter we prefer to illustrate results with SCR = 2 since it provides a better picture of the statistical detector and in particular of its variations (higher *SCR* would have masked it).

Now, the ROC curve is visible in the linear plot showing still adequate performances (Chaney et al. 1990):

$$P_D = 0.75 \text{ with } P_F = 10^{-3},$$
$$P_D = 0.85 \text{ with } P_F = 10^{-2}. \qquad (5.81)$$

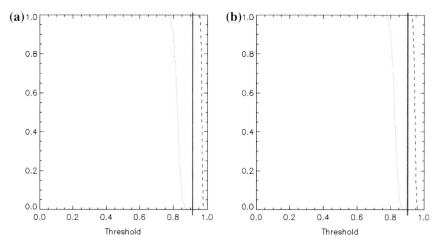

Fig. 5.21 Threshold selection for detection of **a** $SCR = 2$ and **b** $SCR = 1$. *Dotted line*: P_F; *Dashed line*: P_D

5.4.3 Selection of the Threshold

In this section, a practical procedure to select the threshold is proposed. In detection theory, several methodologies were designed to optimise the threshold selection. They mainly concern with the minimisation of P_F and the maximisation of P_D as the Neyman-Pearson or Bayesian methodologies (Kay 1998).

Fortunately, the proposed detector has excellent ROC and a straightforward strategy can be adopted. The idea is to choose the threshold in the region where $P_D \approx 1$ and $P_F \approx 0$. As shown in the previous section, these regions are relatively wide and the choice can be easily made graphically plotting the two probabilities together. Figure 5.21 illustrates an example of this procedure, where the dotted and dashed lines represent respectively P_F and P_D and the red line is the selected threshold.

5.5 Estimation of DF Through Numerical Simulation

In order to validate the derived analytical expression, we performed a series of numerical simulations of the detector starting from the components of the scattering vector: complex Gaussian zero mean. In numerical simulation, the exact *pdf* cannot be achieved, but an approximation regarded as a discrete probability function (DF) $p_\Gamma(\gamma)$ can be obtained (Pearson 1986; Gray and Davisson 2004). As in the previous section, we are interested in the $p_\Gamma(\gamma)$ as a function of the *SCR*. Hence, $p_\Gamma(\gamma|s = SCR)$ will be estimated. As before, the latter is not a conditional probability since the *SCR* is deterministically fixed.

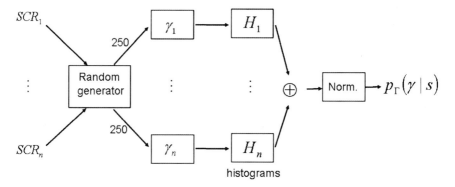

Fig. 5.22 Block diagram for the detector

Fig. 5.23 Detector discrete
probability function (DF).
Simulation: coloured clutter

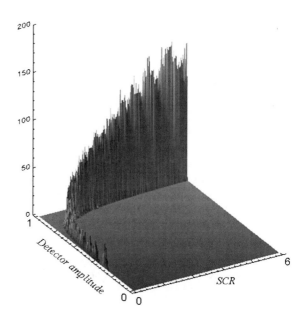

5.5.1 Coloured Clutter Hypothesis

The first hypothesis examines coloured clutter. The block diagram for the generation of the DF is illustrated in Fig. 5.22. The first step defines a set of 250 realisations of the detectors given a fixed *SCR*. The *SCR* are defined increasing the mean of the original random variables used to generate the coherence. In other words, 250 coherences γ_i are generated for each *SCR*. Subsequently, the histogram of the coherences γ_i is calculated for any given *SCR*. This exploits information about the distribution of γ_i.

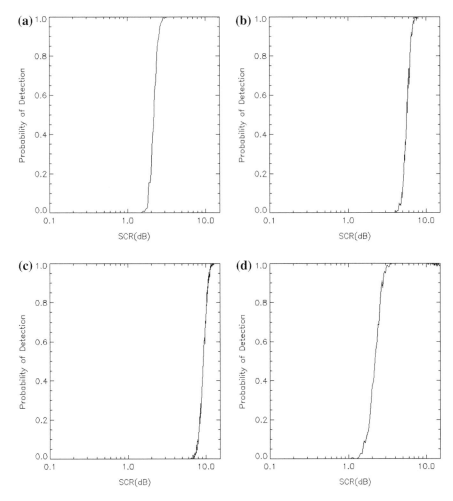

Fig. 5.24 Simulated probabilities of detection P_D against $SCR(\text{dB})$ for different detector parameters (see Table 5.7)

Table 5.7 Detector parameters

	SCR	RedR	Threshold	Window size
(a)	Variable	0.25	0.95	25
(b)	Variable	0.25	0.98	25
(c)	Variable	1	0.95	25
(d)	Variable	0.25	0.95	9

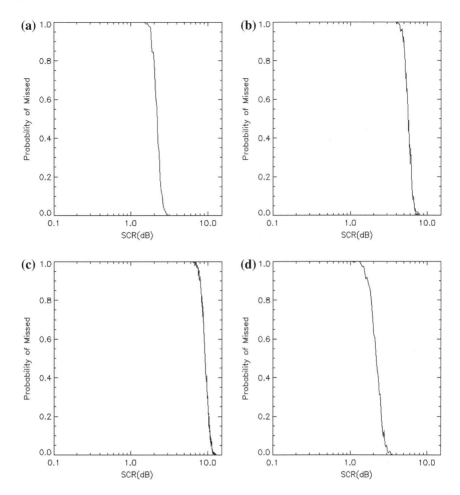

Fig. 5.25 Simulated probabilities of missed detection P_M against SCR(dB) for different detector parameters (see Table 5.7)

The '+' block merges the columns together to form a matrix (i.e. each column is a histogram). The last step performs the normalisation for the columns, since they represent probabilities.

The obtained DF for $RedR = 0.25$ and $N = 25$ is plotted in Fig. 5.23. For low values of coherence the peaks are generally lower showing higher variance of the single column distribution. The general DF trend (Fig. 5.23) is in agreement with the analytical solution (Fig. 5.2) showing the appropriateness of the derived analytical expression. However, the peak values of the probability in the DF could differ from the *pdf* since the DF is equal to the *pdf* only in the limit of infinitesimally small intervals and infinite realisations (Antoniou 2005):

Fig. 5.26 Detector discrete
probability function (DF):
hypothesis 0

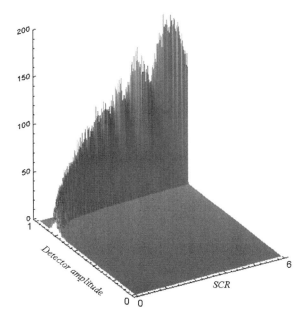

Fig. 5.27 Detector discrete
probability function (DF):
hypothesis 1

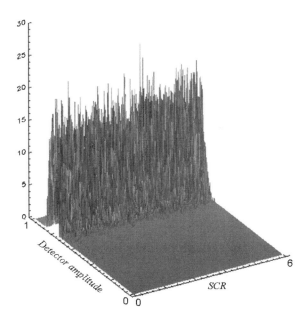

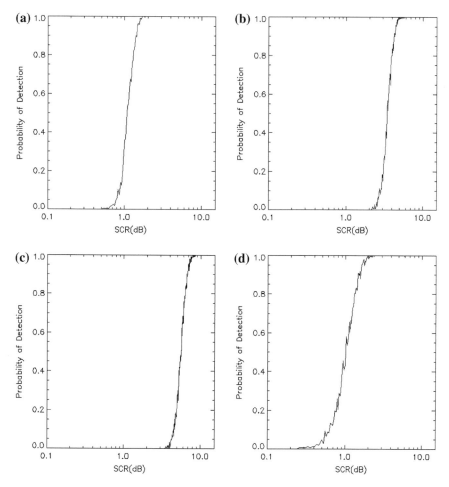

Fig. 5.28 Simulated probability of detection P_D with white clutter against *SCR* for different detector parameters (see Table 5.7)

$$\lim_{\substack{\Delta \to 0 \\ N \to \infty}} DF = pdf, \tag{5.82}$$

where Δ is the interval used to calculate the probabilities and N is the number of realisations.

Once the DF is available this can be used to calculate the characteristic probabilities:

$$P_D(\gamma \geq T | s = SCR_0), \ P_M(\gamma < T | s = SCR_0) \text{ and } P_F(\gamma \geq T | s = 0). \tag{5.83}$$

In Fig. 5.24 the trends of the probabilities of detection P_D are presented. Table 5.7 illustrates the parameters used for the simulations.

Following the same processing the probabilities of missed detection P_M can be estimated (Fig. 5.25 and Table 5.7).

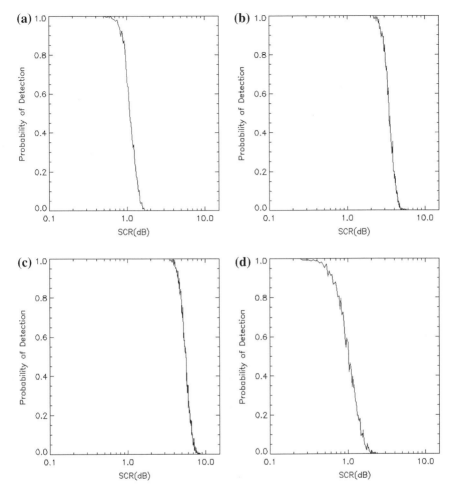

Fig. 5.29 Simulated probability of missed detection for white clutter against SCR(dB) for different detector parameters (see Table 5.7)

The comparison of the simulated and the analytical results show extraordinary agreement (the plots overlaps each other), confirming once again the suitability of the analytical expression.

5.5.2 White Clutter Hypothesis

The more general hypothesis of white clutter is assumed. The simulation adopts exactly the same processing used for coloured clutter except for the addition of one more clutter component summing to the target. Figures 5.26 and 5.27 illustrate the DF as function of the SCR for respectively hypothesis 0 (i.e. target plus clutter) and hypothesis 1 (i.e. only clutter).

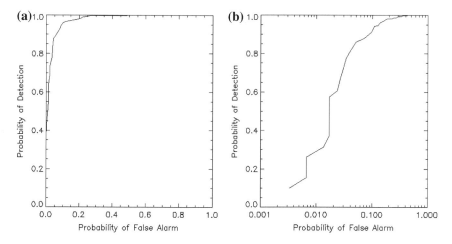

Fig. 5.30 Simulated ROC for $SCR = 0.5$ and window size equal to 9. **a** linear scale, **b** dB scale

The values are the same of Table 5.1 (but variable SCR).

The simulated surface overlaps adequately with the analytical one, showing a characteristic starting point different from zero. Again the peak values are similar but we cannot expect a perfect matching. As for the analytical case, when the target is absent the DF is constant against the SCR, although the power of the clutter is increasing.

The next analysis considers the simulation of the probabilities of detection P_D (Fig. 5.28 and Table 5.6). Again the comparison of simulated and analytic plots reveals great agreement (the plots can be overlapped).

Figure 5.29 presents the probabilities of missed detection P_M (Table 5.6).

The probability of false alarm P_F presents a particular scenario, since the numerical simulation is not able to estimate it properly (or at all). In fact, the probability that some realisations set the detector upper the threshold in absence of a target is extremely small (Papoulis 1965; Monahan 2001). The minimum probability which can be estimated numerically is the reciprocal of the number of samples used

$$P_{\min} = \frac{1}{N}, \tag{5.84}$$

where N is the number of realisations. In these simulations P_F is several orders of magnitude smaller than $P_{\min}$, resulting in improper results or zero (Pearson 1986). The only simulation which was able to return an appreciable P_F was the challenging scenario with small window and weak target.

The last experiment concerns the ROC curves. The simulations were carried out using the same parameters previously exploited for the analytic treatment. Considering the particularly small value of the false alarm, the only ROC which we were able to plot was the last one with window size equal to 9 and $SCR = 0.5$ (Fig. 5.30 and Table 5.6).

References

Antoniou A (2005) Digital signal processing: signals, systems and filters. McGraw Hill, New York

Chaney RD, Bud MC, Novak LM (1990) On the performance of polarimetric target detection algorithms. Aerosp Electron Syst Mag IEEE 5:10–15

Cloude SR (1995) Lie groups in EM wave propagation and scattering, Chap. 3. In: Baum C, Kritikos HN (eds) Electromagnetic symmetry. Taylor and Francis, Washington, pp 91–142, ISBN 1-56032-321-3

Cloude SR, Pottier E (1996) A review of target decomposition theorems in radar polarimetry. IEEE Trans Geosci Remote Sens 34:498–518

Franceschetti G, Lanari R (1999) Synthetic aperture radar processing. CRC Press, Boca Raton

Fung AK, Ulaby FT (1978) A scatter model for leafy vegetation. IEEE Trans Geosci Electron 16:281–286

Gray RM, Davisson LD (2004) An introduction on statistical signal processing. Cambridge University Press, Cambridge

Kay SM (1998) Fundamentals of statistical signal processing, vol 2: detection theory. Prentice Hall, Upper Saddle River

Lee JS, Jurkevich I, Dewaele P, Wambacq P, Oosterlinck A (1994) Speckle filtering of synthetic aperture radar images: a review. Remote Sens Rev 8:313–340

Li J, Zelnio EG (1996) Target detection with synthetic aperture radar. IEEE Trans Aerosp Electron Syst 32:613–627

López-Martínez C, Fàbregas X (2003) Polarimetric SAR speckle noise model. IEEE Trans Geosci Remote Sens 41:2232–2242

Mathews JH, Howell RW (2006) Complex analysis for mathematics and engineering. Jones and Bartlett, London

Monahan JF (2001) Numerical methods of statistics. Cambridge University Press, Cambridge

Novak LM, Owirka GJ, Netishen CM (1993) Performance of a high-resolution polarimetric SAR automatic target recognition system. Linc Lab J 6:11–24

Novak LM, Owirka GJ, Weaver AL (1999) Automatic target recognition using enhanced resolution SAR data. IEEE Trans Aerosp Electron Syst 35:157–175

Oliver C, Quegan S (1998) Understanding synthetic aperture radar images. Artech House, Boston

Papoulis A (1965) Probability, random variables and stochastic processes. McGraw Hill, New York

Pearson CE (1986) Numerical methods in engineering and science. Van Nostrand Reinhold Company, New York

Riley KF, Hobson MP, Bence SJ (2006) Mathematical methods for physics and engineering. Cambridge University Press, Cambridge

Touzi R, Lopes A, Bruniquel J, Vachon PW (1999) Unbiased estimation of the coherence from multi-look SAR data imagery. IEEE Trans Geosci Remote Sens 37:135–139

Treuhaft RN, Siqueria P (2000) Vertical structure of vegetated land surfaces from interferometric and polarimetric radar. Radio Sci 35:141–177

Tsang L, Kong JA, Shin RT (1985) Theory of microwave remote sensing. Wiley Interscience, New York

Chapter 6
Validation with Airborne Data

6.1 Introduction

In the previous section, the statistics of the detector were derived to establish its theoretical performance. Although the assessment returned promising results, the validation on real data is an unavoidable step, since in real scenarios the performance of an algorithm can be dramatically degraded by factors which cannot be easily taken into account in a theoretical model (Campbel 2007).

With the purpose of performing an exhaustive validation, several typologies of targets and datasets will be taken into consideration, exploiting airborne and satellite sensors, ranging among several frequencies and resolutions. In this thesis, the validation process will be divided in two main chapters. The current one is dedicated to the more favourable scenario of airborne data, while the next chapter will treat the more challenging scenario of satellite data (Campbel 2007).

In the first part of the validation, the presence of standard targets (as explained in the previous section) will be investigated. They represent an interesting starting point due to their easy association with real targets. In the second part, general targets will be explored and the best single target to focus for the detection of the corresponding real target will be examined. We want to stress that this thesis does not present an ad hoc study for a specific real target (e.g. a specific car with a specific orientation with respect to flight direction), since we intend to present a general detector. The detection and identification of the backscattering from a particular object goes outside the aim of this thesis.

A. Marino, *A New Target Detector Based on Geometrical Perturbation Filters for Polarimetric Synthetic Aperture Radar (POL-SAR)*, Springer Theses, DOI: 10.1007/978-3-642-27163-2_6, © Springer-Verlag Berlin Heidelberg 2012

6.2 Presentation of the E-SAR Data and General Considerations

In this first series of experiments the detector is applied on a quad polarimetric (i.e. *HH*, *VV*, *HV* and *VH*) L-band SAR dataset. L-band is relevant to target detection for its ability to penetrate foliage (FOLPEN) (Fleischman et al. 1996). The datasets were acquired by the DLR (German Aerospace Centre) during the SARTOM campaign in 2006 (Horn et al. 2006) with the E-SAR airborne system. A noteworthy characteristic of the E-SAR radar sensor is its spatial resolution: 2.2 m in range and 0.9 m in azimuth. As shown in the previous section, the theoretical detection performance improves with increasing size of the averaging window employed to estimate the coherence. The drawback is the loss of resolution. For this reason, a high resolution sensor allows sufficient averaging with adequate final resolution. Considering a 5×5 moving window is employed, the final averaged cell will be 11×4.5 m which is sufficient for vehicles and small buildings (Novak et al. 1999).

The SARTOM campaign was designed to put under test the detection capabilities of the advanced SAR techniques of tomography and polarimetry. For this purpose, a set of artificial targets were deployed in open fields and under canopy cover, making this dataset particularly suitable for our experiments (Horn et al. 2006).

Figure 6.1 presents the aerial photograph of the test site (Google Earth), with markers for the location of targets in the scene. Figure 6.2 is a colour composite RGB image where the three colours are the components of the Pauli vector (*Red*: *HH* − *VV*, *Green*: *2HV*, *Blue*: *HH* + *VV*) (Cloude 2009; Lee and Pottier 2009; Mott 2007; Ulaby and Elachi 1990; Zebker and Van Zyl 1991).

Comparing the radar image with the aerial photograph, the geometrical distortions affecting the radar image are evident. In particular, the radar image is compressed along the range direction since the azimuth resolution is higher (Horn et al. 2006).

The brightest regions in our SAR image correspond mainly to forests. The brightness is due to the presence of several scatterers (e.g. branches, leaves, etc.) with dimensions comparable to the wavelength (Attema and Ulaby 1978; Durden et al. 1999; Fung and Ulaby 1978; Lang 1981; Treuhaft and Cloude 1999; Treuhaft and Siqueria 2000; Tsang et al. 1985; Woodhouse 2006). The multi scattering generated by the elements is consistent and relatively isotropic (i.e. in all the directions), consequently a significant fraction of energy is scattered backward (i.e. toward the receiving antenna). Polarimetrically, the tree crown can be modelled as a statistical volume composed by several oblate particles with or without a preferential orientation (Fung and Ulaby 1978; Treuhaft and Cloude 1999; Treuhaft and Siqueria 2000).

In the literature, different models were developed to describe volume scattering, one of the most common is the Random Volume (RV), which considers particles (i.e. spheres, or dipoles) randomly oriented, or the Oriented Volume, where the particles have a preferential orientation.

Fig. 6.1 Aerial photograph of the test area (Google Earth). *CR*: trihedral corner reflector; *WOLF*: jeep; *ILTIS*: jeep covered by net; *LKW*: truck with container on the top; *PANZER*: tanks

The backscattering from bare ground is generally less bright. It can be modelled as a rough surface (i.e. Bragg scattering) where most of the energy is scattered forward (Cloude 2009). If the roughness is very small compared with the wavelength then most of the energy of the incident wave (which is not absorbed or refracted inside) will be reflected in the mirror direction following the Fresnel laws (Stratton 1941; Rothwell and Cloud 2001). In some limits, the roughness (compared with the wavelength) is directly related with the energy scattered back to the sensor. Polarimetrically, bare ground behaves like a surface, hence the odd-bounce represented by the first component of the Pauli scattering vector (i.e. $HH + VV$) is particularly strong (Cloude 2009; Hajnsek et al. 2007). A way to understand the scattering from a rough dielectric surface is to consider an ideal surface and start making it progressively less ideal.

(a) An infinite metallic smooth surface scatters only in the foreword direction and *HH* and *VV* have exactly the same amplitude but opposite phase (as long as the incident waves are in phase). This is because the reflection of the vertical

Fig. 6.2 L-band RGB Pauli composite image of the test area. *Red*: *HH* − *VV*, *Green*: *2HV*, *Blue*: *HH* + *VV*

component changes sign. However with the BSA coordinate system the change in the direction of the horizontal axis results in a double change of sign making *HH* and *VV* in phase (Cloude 1987; Huynen 1970).

(b) An infinite dielectric smooth surface will still scatter only in the foreword direction but now *HH* is not equal in amplitude to *VV*, due to the Brewster angle which will make the *VV* component smaller than the *HH* one. As in the metallic case, the surface does not introduce depolarisation (Cloude 2009; Rothwell and Cloud 2001).

(c) The introduction of surface roughness generates spreading of the scattered energy away from the foreword direction (covering the backward direction as well).

Finally, in the case of bare ground, the surface is dielectric and rough. The roughness of the surface produces depolarisation, hence *HV* and *VH* are not zero anymore, and *HH* and *VV* are not exactly in phase. For Bragg scattering, in backscattering the balance between *VV* and *HH* reverses and *VV* is higher than *HH*. The effects of the Brewster angle on target detection of multiple reflections with the ground surface will be presented in more detail in the next section.

The artificial targets (i.e. corner reflectors, containers, and vehicles) deployed in open field are rather evident in Fig. 6.2, since they are generally bright. The brightness is mainly associated with their geometrical shape which is favourable to the formation of mirrors or corners (Curlander and McDonough 1991; Li and Zelnio 1996; Novak et al. 1999). Evidently, the intensity of the backscattering depends on the dimensions of the corner and its radar cross section can be calculated as shown in the first chapter. The connection between artificial targets and metallic corners is the focal idea of detectors which set thresholds on the amplitude of the backscattering in the linear co-polarisations *HH* or *VV* (as presented in the third chapter). A slightly more refined approach considers thresholds on the first two elements of the Pauli scattering matrix (namely odd-bounce and even-bounce).

In Fig. 6.2, the features appearing as geometrical shapes in open field are metallic nets (please note they are not marked in Fig. 6.1).

Regarding the targets deployed under forest canopy, they are generally not visible and separable by the surrounding clutter in the RGB image (Fleischman et al. 1996; Cloude et al. 2004). This is due to two main reasons:

(a) Microwave radiation is able to penetrate dielectric mediums, with penetration depth related to the permittivity of the medium (associated with the density in the case of cluster medium) (Rothwell and Cloud 2001; Stratton 1941). A tree canopy is composed by several particles separated by air gaps (which occupy most of the volume). The canopy can be penetrated by the EM radiation in L-band, but it suffers attenuation due to particle absorption and dispersion (which scatters the energy in different directions). Consequently, the amount of energy able to reach the ground beneath the canopy is merely a fraction of the incident one (Fung and Ulaby 1978; Treuhaft and Siqueria 2000; Tsang et al. 1985). Once reached the target, the wave has to travel back toward the sensor along the same canopy path. The two way attenuation of the canopy can drastically lower the backscattering from the target.
(b) In a forested area the surrounding clutter is much higher than bare ground, since the forest has high backscattering (as explained previously) (Kay 1998).

In conclusion, the power backscattered by the target under canopy cover is reduced while the power of the surrounding clutter increases. In some instances, this leads to a target backscattering lower than the background making unfeasible the detection based solely on the backscattering amplitude.

In a first experiment, the presence of multiple reflections will be investigated in order to detect vehicles and corner reflectors. Additionally, oriented dipoles will be explored in order to detect wires. In the second part, general targets will be investigated, with the purpose of examining the best single target to use for the detection of the corresponding real target.

6.3 Standard Target Detection

As mentioned previously, artificial targets are primarily composed of basic shapes and corners, which could be selected in a first attempt. In Chap. 4, the polarimetric characteristics of standard targets were examined, here, an expected real target will be associated with the theoretical one (Cloude 2009; Lee and Pottier 2009). The detection will be aimed at:

(a) Odd bounce: these are corners with metallic planes where the wave has been reflected an odd number of times before being redirected to the sensor. Examples of this typology are surfaces facing the sensor and trihedral corners.

(b) Even bounce (horizontal): these are again corners with metallic planes where the wave suffers an even number of reflections before reaching the sensor. In particular, the horizontal orientation of the corner line is an important specification, since the target is not invariant to rotation along the Line of Sight (LOS). Examples of even bounce are dihedral corners like walls or vehicles oriented along the azimuth.

(c) Horizontal dipole: a dipole is generated by a line of current. It will scatter a linear polarisation (i.e. zero ellipticity) with orientation equal to the wire. Again, the orientation of the dipole is an important specification. Examples are wires along the azimuth direction and parallel to the ground, but also narrow cylinders, like long thin branches.

(d) Vertical dipole: same as horizontal dipole but oriented along the vertical direction. By vertical direction we mean any direction on the plane passing through the range direction and orthogonal to the horizontal plane. Although any wire on this plane will be interpreted as vertical dipole, the amount of backscattering coming from the target is a function of its orientation on this plane (since a dipole is not isotropic along the direction of its axis). For instance the return is particularly strong when the wire is perpendicular to the range direction. Another occurrence of vertical dipoles is when the wire is normal to the horizontal ground plane, since it produces double bounces with the ground (as long as the latter is sufficiently smooth and flat).

Standard targets represent ideal metallic targets, however real targets will generally be slightly dissimilar from them. From a geometrical point of view, a real single target can still be represented by a vector in the target space since it is deterministic and coherent. However, as a consequence of the non-ideal nature of the target, its vector will be slightly different from the ideal one (Cloude 1995b). This difference can be described as an angular distance between the two vectors (Rose 2002; Strang 1988). In the detector, when the perturbation of the ideal target $\underline{\omega}_T$ is performed (and the threshold is set), the detection is restricted to a defined cone of vectors with the target to be detected as the axis (a range of angular distances from the target). If the real target is inside the cone of detection it will still be detected, otherwise a different typology of single target must be exploited for the real target. In this context, the perturbation (and the set of the threshold) is

Table 6.1 Parameters used for the detector

Window	Area size (Rg, Az)	T reflections	T dipoles	RedR
5 × 5	250 × 250 pixels	0.97	0.95	0.25

an instrument to adjust the angle variation from the ideal target which is assumed as acceptable (as the dispersion equation presented in Appendix 2).

The results of the detection are masks. When a target triggers the detection because the coherence between ω_T and ω_P is above the threshold, the mask records the value of the coherence scaled linearly between the threshold and one. In this way, a measure of the dominance of the target based on the coherence amplitude is assigned (Kay 1998). The mask will be:

$$
\begin{aligned}
m(rg, az) &= 0, \quad \text{if } \gamma_d < T, \\
m(rg, az) &= \gamma_d, \quad \text{if } \gamma_d \geq T,
\end{aligned}
\tag{6.1}
$$

where rg stands for range, az for azimuth, γ_d is the detector as presented in the previous chapter and T the threshold.

Table 6.1 shows the main parameters selected in the detection. A clarification must be provided regarding the selection of the two threshold values. In the previous chapter, a methodology to select the threshold was developed and tested for $SCR = 2$ and $SCR = 1$. These two examples where selected because they present a favourable didactic picture of the statistical detector. However, in a practical detection we could be not interested in targets with $SCR = 2$ since the dispersion equation shows a relatively large variation of the target (35 degrees of angular variation) with possible detection of similar coherent targets. For this reason, the thresholds used were optimised for white clutter and $SCR = 4$ and $SCR = 6$. These two values of SCR were not treated in the statistical evaluation (previous chapter) since with them the detector presents a strong deterministic behaviour and the variation effects are not clearly visible.

Moreover, the reason for two different thresholds for reflections and dipoles is related to the brightness of multiple reflections in open field. This makes their return particularly dominant on the surrounding clutter and a higher threshold is to be preferred in the case of high SCR (signal to clutter ratio) since it reduces the false alarm rate. In conclusion, if we have a priori information about the typology of scenario (e.g. open field) a more clever selection of the threshold can be performed. The dipoles are generally not particularly bright, hence a standard threshold (i.e. SCR) must be chosen.

Figure 6.3 presents the results of the detection of multiple reflections and oriented dipoles. Besides, Fig. 6.4 illustrates the photographs of some detected targets taken during a survey of the test area. The L-band RGB Pauli image Fig. 6.3a is given as comparison with markers in order to identify targets of interest (250 × 250 pixels). A jeep is deployed in the middle of the image (Mercedes Benz 250 GD, also named 'Wolf') and the two bright points above and below the jeep are trihedral corner reflectors positioned for calibrations (top 149 cm; bottom 70 cm).

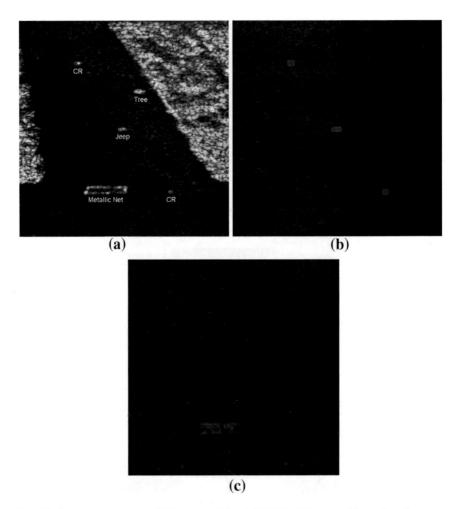

Fig. 6.3 Detection over open field area. **a** L-band RGB Pauli image with markers for some targets. *Red*: *HH* − *VV*, *Green*: *2HV*, *Blue*: *HH* + *VV*. **b** RGB mask for multiple reflection. *Red*: Even-bounce, *Green*: zero, *Blue*: Odd-bounce; **c** RGB mask for oriented dipole detector. *Red*: horizontal dipole, *Green*: zero, *Blue*: vertical dipole. The intensity of the masks is related to the detector amplitude

Finally, on the bottom of the image there is a vertical metallic net (they were used to delimit areas). The range direction is along the vertical axes (bottom to top) (Horn et al. 2006).

Any masks display the information about two different detections gathered together in an RGB composite colour image. The masks are:

(a) Multiple reflections (Fig. 6.3b): targets constituted of metallic surfaces and corners are represented in the same mask. Red stands for even bounce and Blue for odd bounce (as in the Pauli basis convention).

(b) Oriented dipoles (Fig. 6.3c): the wire or narrow cylinders can be visualised on this mask. Red represents horizontal and blue vertical dipoles (as in the Lexicographic basis convention).

The algorithm correctly detects the trihedral corner reflectors (CR) as a source of odd-bounces (blue spots on the mask). The return from the CRs is especially pure and strong since the faces are metallic. The diffraction on the edges is a factor which can reduce slightly the purity but at L-band this is rather negligible (Rothwell and Cloud 2001). The jeep presents mainly even bounces, presumably due to the double bounce reflections between ground and vertical surfaces of the jeep (and vice versa). Additionally, some even bounce targets are located on the forest edge, due to the trunk-ground double bounce which is particularly strong at the forest edge where the trunk plane is not shadowed by canopy or other trunks. These typologies of targets (except the CRs) are less pure since at least one of the planes (i.e. the ground) is dielectric. Furthermore, the planes are rough surfaces, the angle between them can be slightly different from normal and the orientation of the corner line could not be exactly horizontal (e.g. presence of slopes). In the next section a procedure is described to focus the detector more sharply on the real target of interest.

Regarding the oriented dipole, the targets composed by reflecting surfaces (especially the CRs) disappear completely. The metallic net (Fig. 6.4b) is detected as a horizontal dipole, since the horizontal wires scatter more compared to the vertical ones due to the radar geometry (as explained previously).

It is interesting to note that the same polarimetric behaviour is shared by all the metallic nets present in the dataset, hence horizontal dipoles can be reliably exploited for their detection. The isolated tree at the edge of the forest is detected as a horizontal dipole due to its long and thin horizontal branches (as visible in Fig. 6.4c). These branches are long cylinders with diameter of one or two centimetres, hence they are interpreted as narrow cylinders by the 24 cm wavelength (Cloude 1995a, 2009).

Several vertical dipoles are detected on the ground. A survey of the test area revealed the presence of bushes constituted of big wooden vertical stems with height around one and half meters (Fig. 6.4d).

The next detection exercise is aimed to test the algorithm performance in a more challenging environment, when the target is deployed beneath canopy cover (Fig. 6.5). In these circumstances both the thresholds for multiple reflections and oriented dipoles are set to 0.95, since the targets are not expected to be particularly dominant and the rejection of clutter is a primary issue. In general, forest clutter should not amplify dramatically the false alarm rate since volume contribution (specially randomly oriented particles) is spread over all the target space resulting in confusing the polarimetric information (i.e. making the entropy higher) and reducing the probability that the detector surpasses the threshold (Tsang et al. 1985).

The deployed targets are three trihedral corner reflectors (top: 149 cm, bottom left: 70 cm, bottom right: 90 cm). In the RGB Pauli image (Fig. 6.5a) the CRs are

Fig. 6.4 Photographs of some detected targets (Courtesy of DLR) **a** Wolf1. **b** Metallic net. **c** Sparse tree with horizontal branches. **d** Bushes with vertical wooden stems

not recognisable, since the surrounding clutter (i.e. forest) has a bright return masking them out. Conversely, they are easily detected as source of odd bounce in the multiple reflections mask (Fig. 6.5b). Additionally, the algorithm is able to detect bare ground in the upper part of the image. As mentioned before, the ground return can be modelled as a Bragg surface (Cloude 2009). In this last exercise, we are able to detect bare ground because the threshold is lower than before and weak targets can be detected.

Regarding the even bounce, several trunk-ground double bounces can be identified, especially in proximity of the forest clearing separating the top and bottom *CRs* (i.e. darker line running along the azimuth). When the forest density is lower, the trunk surface has more probability to generate a dihedral with the ground.

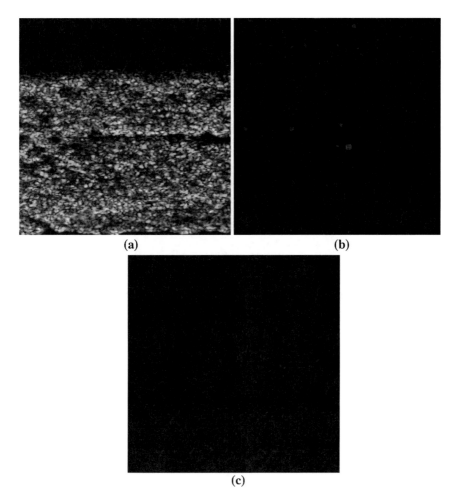

(a)   (b)

(c)

Fig. 6.5 Detection over forested area: corner reflectors. **a** L-band RGB Pauli image. *Red*: $HH - VV$, *Green*: $2HV$, *Blue*: $HH + VV$. **b** RGB mask for multiple reflection. *Red*: Even-bounce, *Green*: zero, *Blue*: Odd-bounce. **c** RGB mask for oriented dipole detector. *Red*: horizontal dipole, *Green*: zero, *Blue*: vertical dipole. The intensity of the masks is related to the detector amplitude

Finally, it is not possible to detect dipoles with marked preferential orientation in the forest (Fig. 6.5c). This is in line with the RVoG model for L-band, where the forest structures are random and do not present particular orientations (Fung and Ulaby 1978; Treuhaft and Siqueria 2000; Papathanassiou and Cloude 2001). Regarding the vertical dipoles, the ground does not present detections anymore, in fact it is detected as single bounce (the field visit confirmed the absence of bushes in that area). The photograph image of two of the three corner reflectors (two on the bottom) is presented in Fig. 6.6.

In order to offer another example of detection under foliage, Fig. 6.7 depicts the detections where a different typology of target is investigated: a 20 ft steel

Fig. 6.6 Photograph of two of the three trihedral corner reflectors in the forest (Courtesy of DLR)

container was deployed on a forest clearing as shown in Fig. 6.8b. The side-looking arrangement of the radar makes the container completely under cover of the forest canopy (on the left hand side). The photograph also gives an idea about the density of the forest. Figure 6.7b reveals clearly the detection of the container as source of even bounces (i.e. a metallic wall over a ground surface). In this fortunate case, the backscattering from the container is relatively bright, consequently we could use the higher threshold 0.97 for multiple reflections. This makes the final mask cleaner, rejecting all the weak and non ideal targets (e.g. bare ground or trunks).

Regarding the odd bounces, there are two fascinating features detected in proximity of the bottom left corner of the forest stand. The stripe shaped is a slope used for training tank drivers (Fig. 6.8c). The main slope faces the sensor acting like a mirror (i.e. single bounce) although it is not metallic. The other blue spot is a bunker as shown in Fig. 6.8d. Here, the side facing the sensor is the one inside the hole and the frontal prominent edge.

As expected, horizontal branches on the forest edge generates horizontal dipoles. Regarding the detected point on the lower left corner of the forest stand, they correspond to a military observation turret, as shown in Fig. 6.8a.

The turret is a noteworthy example of a complex target. It is detected as a horizontal dipole, hence its *HH* component (or something close to it) is dominant against all the others components. However, the interpretation of the peculiar interaction between the turret and the incident wave cannot be trivially derived from a mere visual inspection of the target (Stratton 1941). This is one of the

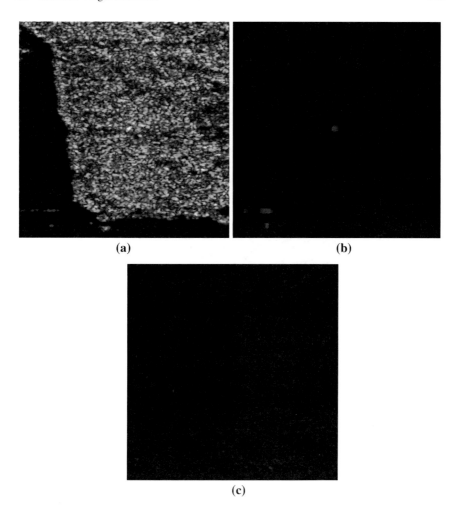

Fig. 6.7 Detection over forested area: container. **a** L-band RGB Pauli image. *Red*: *HH − VV*, *Green*: *2HV*, *Blue*: *HH + VV*. **b** RGB mask for multiple reflection. *Red*: Even-bounce, *Green*: zero, *Blue*: Odd-bounce; **c** RGB mask for oriented dipole detector. *Red*: horizontal dipole, *Green*: zero, *Blue*: vertical dipole. The intensity of the masks is related to the detector amplitude

reasons for the choice to test the detector with standard targets, since they generally have an easy physical counterpart (except few exceptions, as in this case of the turret). We left out other theoretical targets (e.g. helixes) since we expect a more complicated association with real targets based on a visual inspection. This would have the consequence of adding ambiguity to the validation (Cloude 2009; Rothwell and Cloud 2001; Stratton 1941). Clearly, if the exact signature of the target of interest is known the detector can be focused on it (since we know that the theoretical target is associated with our real target of interest).

Fig. 6.8 Photograph pictures of some detected targets (Courtesy of DLR). **a** Military observation tower. **b** Container in forest. **c** Tank training slope. **d** Bunker

Finally, there are some vertical dipoles on the ground due to bushes, but not as much as in the previous example.

In the last experiment with E-SAR data, a different typology of targets is under examination. The entire mathematical derivation is based on the possibility of characterising the polarimetric behaviour of the target with a single scattering vector (since it is a single target). In other words, the target must be polarimetrically stable or deterministic, which commonly translates into point targets or targets composed of a stable compound of objects (with the same polarimetric behaviour) in the resolution cell. Single targets are often associated (and sometimes confused) with coherent targets which are targets that do not present speckle variation, because they are deterministic. As explained in the first chapter, speckle is due to the coherent sum of several scatterers in the same resolution cell with different possible combinations.

Even though the correspondence between single and coherent target is commonly acceptable in practical cases, there is a theoretical difference which sometimes can be observed on radar data (Dong and Forster 1996). The provided example is a target distributed over several pixels (i.e. composed by several scatterers) where all the scatterers have exactly the same polarimetric response. Such a target can still be characterised with a single scattering matrix, even though it is distributed. In other words, if in n neighbour pixels the scattering matrix has the same polarimetric information but different overall amplitude and phase (due to amount of scatterers and location), after the averaging the scattering matrix becomes:

$$\kappa \begin{bmatrix} a & c \\ c & b \end{bmatrix} = \frac{1}{n} \left(C_1 e^{j\phi_1} \begin{bmatrix} a & c \\ c & b \end{bmatrix} + \ldots + C_n e^{j\phi_n} \begin{bmatrix} a & c \\ c & b \end{bmatrix} \right). \tag{6.2}$$

In conclusion, the scattering matrix of the total target is unchanged except for a multiplicative complex factor (which can be neglected in the polarimetric context). Clearly, although the use of the scattering matrix is visually effective it is not formally completely correct to prove this property since we should demonstrate it on the covariance matrix. In order to have a more rigorous demonstration we present the following proof.

If the covariance matrix is obtained from the Lexicographic scattering vector, the first element of the matrix will be:

$$\begin{aligned}
\left\langle |a|^2 \right\rangle &= \frac{1}{n} \left[\left(aC_1 e^{j\phi_1} + \ldots + aC_n e^{j\phi_n} \right) \left(aC_1 e^{j\phi_1} + \ldots + aC_n e^{j\phi_n} \right)^* \right] \\
&= \frac{1}{n} \left[a \left(C_1 e^{j\phi_1} + \ldots + C_n e^{j\phi_n} \right) a^* \left(C_1 e^{j\phi_1} + \ldots + C_n e^{j\phi_n} \right)^* \right] \\
&= \frac{|a|^2}{n} \left[\left(C_1 e^{j\phi_1} + \ldots + C_n e^{j\phi_n} \right) \left(C_1 e^{j\phi_1} + \ldots + C_n e^{j\phi_n} \right)^* \right] = \kappa \frac{|a|^2}{n}.
\end{aligned} \tag{6.3}$$

The same procedure can be applied to all the other diagonal terms resulting in

$$\left\langle |a|^2 \right\rangle = \frac{\kappa}{n} |a|^2, \quad \left\langle |b|^2 \right\rangle = \frac{\kappa}{n} |b|^2, \quad \text{and} \quad \left\langle |c|^2 \right\rangle = \frac{\kappa}{n} |c|^2. \tag{6.4}$$

Regarding the cross components, we can write

$$\begin{aligned}
\left\langle ab^* \right\rangle &= \frac{1}{n} \left[\left(aC_1 e^{j\phi_1} + \ldots + aC_n e^{j\phi_n} \right) \left(bC_1 e^{j\phi_1} + \ldots + bC_n e^{j\phi_n} \right)^* \right] \\
&= \frac{1}{n} \left[a \left(C_1 e^{j\phi_1} + \ldots + C_n e^{j\phi_n} \right) b^* \left(C_1 e^{j\phi_1} + \ldots + C_n e^{j\phi_n} \right)^* \right] \\
&= \frac{ab^*}{n} \left[\left(C_1 e^{j\phi_1} + \ldots + C_n e^{j\phi_n} \right) \left(C_1 e^{j\phi_1} + \ldots + C_n e^{j\phi_n} \right)^* \right] = \kappa \frac{ab^*}{n}.
\end{aligned} \tag{6.5}$$

which translates in:

$$\left\langle ab^* \right\rangle = \frac{\kappa}{n} ab^*, \quad \left\langle ac^* \right\rangle = \frac{\kappa}{n} ac^*, \quad \text{and} \quad \left\langle cb^* \right\rangle = \frac{\kappa}{n} cb^*. \tag{6.6}$$

Summarising, the operation can be seen as

$$[C] = \frac{\kappa}{n}[C_s], \tag{6.7}$$

which is the multiplication of the covariance matrix for a complex scalar (which does not change the polarimetric characteristics of the target). From the geometrical point of view the final covariance matrix can be obtained from the same scattering vector (the one of the scatterers), consequently it is of rank one and represents a single target. Such a target presents speckle due to its distributed structure but can still be detected due to it polarimetrically single nature.

In order to test the detection for distributed single targets, we focused the algorithm on agricultural areas (Fig. 6.9). Figure 6.9d shows the aerial photograph (Google Earth) of the area, where the radar scene is contained between the two black lines. Here, three fields with stripe shapes are visible between two forest stands. In order to facilitate the interpretation, in the detection masks, the lines indicate the edge of the forest.

The mask reveals that the middle field is dominated by double bounces. Additionally, weak vertical dipoles are detected on the same area. Considering that the value of the backscattering is low (Fig. 6.9a) the target in the cell is expected to have a small radar cross section. A likely real target should resemble a collection of small scatterers with vertical preferential orientation which are able to generate double bounces with the ground. For instance, it could be composed of vertical stems, similar to vertical dipoles positioned close to each other in order to generate double bounces with the ground. Unfortunately, we do not have any picture of the field during the time of acquisition.

6.4 Selectable Detection

The previous section was focused on detection of standard single targets, while here the complete potentiality of the algorithm will be tested by investigating a wider collection of single targets. The exercise is accomplished by modifying the target of interest by gradually rotating the scattering mechanism and examining the variation in the detection mask for the same scene. Such processing could be interpreted as a sensitivity analysis of the real target in polarimetric space. This is aimed to obtain the best single target for the detection of the real one.

In order to rotate the target vector, a parameterisation must be exploited. We opted for the same representations introduced previously for the perturbation analysis:

(a) Huynen parameters (Huynen 1970):

$$[S] = [R(\psi_m)][T(\tau_m)][S_d][T(\tau_m)][R(-\psi_m)], \tag{6.8}$$

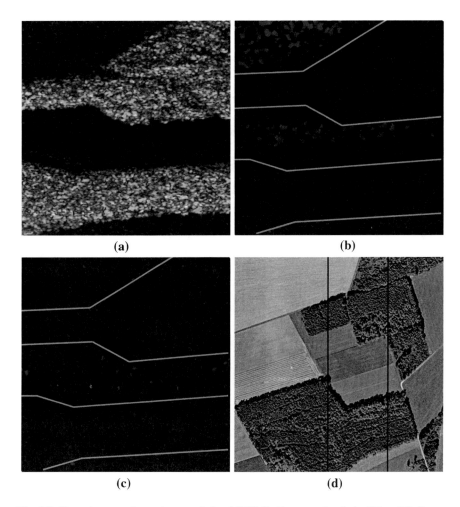

Fig. 6.9 Detection over forested area. **a** L-band RGB Pauli composite. *Red*: *HH* − *VV*, *Green*: 2*HV*, *Blue*: *HH* + *VV*. **b** RGB mask for multiple reflection. *Red*: Even-bounce, *Green*: zero, *Blue*: Odd-bounce. **c** RGB mask for oriented dipole detector. *Red*: horizontal dipole, *Green*: zero, *Blue*: vertical dipole. The *lines in the masks* represent the edge of the forest. The intensity of the masks is related to the detector amplitude. **d** Google aerial photograph of the area

$$[S_d] = \begin{pmatrix} me^{i(\upsilon+\zeta)} & 0 \\ 0 & m\tan(\gamma)e^{-i(\upsilon-\zeta)} \end{pmatrix},$$

$$[T(\tau_m)] = \begin{pmatrix} \cos\tau_m & -i\sin\tau_m \\ -i\sin\tau_m & \cos\tau_m \end{pmatrix},$$

Fig. 6.10 Photographs pictures of some detected targets (Courtesy of DLR). **a** Container. **b** Truck

$$[R(\vartheta_m)] = \begin{pmatrix} \cos \vartheta_m & -\sin \vartheta_m \\ \sin \vartheta_m & \cos \vartheta_m \end{pmatrix},$$

where ϑ_m and τ_m are orientation angle and ellipticity angle of the first eigenvalue (Cros-pol Null), and m, υ, γ and ζ are respectively, target magnitude, target skip angle, characteristic angle and absolute phase. The latter is generally not usable in single pass polarimetry since it is not separable from the phase term due to the distance. Additionally, we generate scattering mechanisms (i.e. unitary vectors), hence $m = 1$. In conclusion, the number of useful Huynen parameters is four.

(b) α model (Cloude 2009; Lee and Pottier 2009):

The scattering mechanism can be represented as

$$\underline{\omega} = \left[\cos \alpha, \sin \alpha \cos \beta e^{i\varepsilon}, \sin \alpha \sin \beta e^{i\mu} \right]^T. \tag{6.9}$$

where α is the characteristic angle (different from γ) and β is twice the target orientation angle. Please note, also in this case the scattering mechanism can be described by four parameters (Fig. 6.10).

The sensitivity analysis is carried out by fixing three of the four parameters and letting the last vary in its entire range of definition. We decided to perform the detection on an area slightly larger (400×400 pixels) and highly populated with artificial targets in order to provide a relatively large picture. The results obtained using these parameterisations can be compared with the standard targets, examining the consistency of the results.

In Fig. 6.11, a sensitivity analysis is performed with the α model, where α is varied in the range $\alpha \in \left[0, \frac{\pi}{2}\right]$, and $\beta = \varepsilon = \mu = 0$. In term of targets, $\alpha = 0$ and $\beta = \varepsilon = \mu = 0$ represents single reflection or isotropic targets, hence in Fig. 6.11b odd bounces are detected, such as corner reflectors and some bare ground. $\alpha = \pi/4$ represents dipoles and $\beta = 0$ establishes they are horizontal, hence the metallic net as well as some branches at forest edge are identified. $\alpha = \pi/2$ is for even bounce

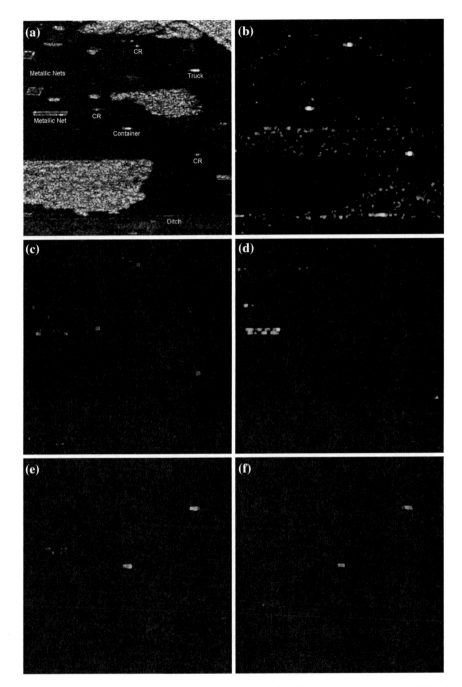

Fig. 6.11 Selectable detection over open field area. $\alpha \in [0, \pi/2]$ and $\beta = \varepsilon = \mu = 0$. **a** L-band RGB Pauli composite image with markers for some targets; **b** $\alpha = 0$; **c** $\alpha = \pi/8$; **d** $\alpha = \pi/4$; **e** $\alpha = 3\pi/8$; **f** $\alpha = \pi/2$

which in the image are generated by a truck and a container (Fig. 6.10 shows their pictures).

$\alpha = \pi/8$ and $\alpha = 3\pi/8$ represents intermediate typologies of targets. This is the first experiment where they are considered, therefore it is appropriate to try to describe them in more detail. In the sensitivity analysis, $\alpha = 3\pi/8$ seems to be able to detect dihedrals constituted of a metallic wall over ground surface better than the standard even bounce. For instance, the container is observed by the sensor as a metallic wall which generates a dihedral with a dielectric surface (i.e. the ground). The ground introduces the Brewster (or pseudo Brewster) angle for the vertical linear polarisation (Rothwell and Cloud 2001; Stratton 1941). Therefore, the backscattering of the *VV* polarisation is lower than in *HH*. This additional contribution of the *HH* polarisation can be interpreted as a coherent horizontal dipole in phase with the *HH* component of the even bounce. Therefore, the final target will be something in between the ideal even bounce ($\alpha = \pi/2$) and the horizontal dipole ($\alpha = \pi/4$).

In the literature this typology of target is regarded as *narrow dihedral* (Cameron 1996). In conclusion, in the case horizontal dihedral with the ground surface are investigated the detector should be focused on narrow dihedrals more than ideal dihedrals.

An equivalent representation to the α model is the Huynen parameterisation. In Fig. 6.12, an exercise is accomplished varying the characteristic angle γ in the range $\gamma \in \left[0, \dfrac{\pi}{4}\right]$.

The characteristic angle is related with the ratio of the two Cross-pol Nulls, where for $\gamma = 0$ only one Cross-pol Null is different from zero (one eigen-value) and for $\gamma = \pi/4$ the two Cross-pol Nulls are equal (i.e. multiple reflections). The other parameters are: $\psi_m = \tau_m = 0$ and $\upsilon = \pi/2$. For $\gamma = 0$ the target selected is a horizontal dipole (since the orientation ψ_m and the ellipticity τ_m are zero). Please note, Fig. 6.12a presents the same detection mask obtained in the previous exercise. On the other hand, for $\gamma = \pi/4$ the target is a multiple reflection and $\upsilon = \pi/2$ states it is an even bounce (the two eigenvalues have opposite phase). In Fig. 6.12f shows the expected even bounce detection with the container and truck. The intermediate masks detect targets in between the extreme cases. The metallic net seems to be rather persistent in the detections appearing in several masks but this could just be related with the non linearity of the scattering mechanism dependence. Moreover, again the container and truck seem to be better detected when the algorithm is focused on a combination of horizontal dipole and ideal dihedral rather than the ideal dihedral alone.

The last sensitivity analysis in Fig. 6.13 considers the variation of the skip angle υ in $\upsilon \in [0, \pi/2]$. The latter is related to the phase difference between the two Cross-pol Nulls. The parameters set here are $\gamma = \pi/4$ (i.e. multiple reflection) and $\psi_m = \tau_m = 0$, consequently the detection sweeps from odd to even bounce. The images at the extreme of the range values seem to be in agreement with the previous detections.

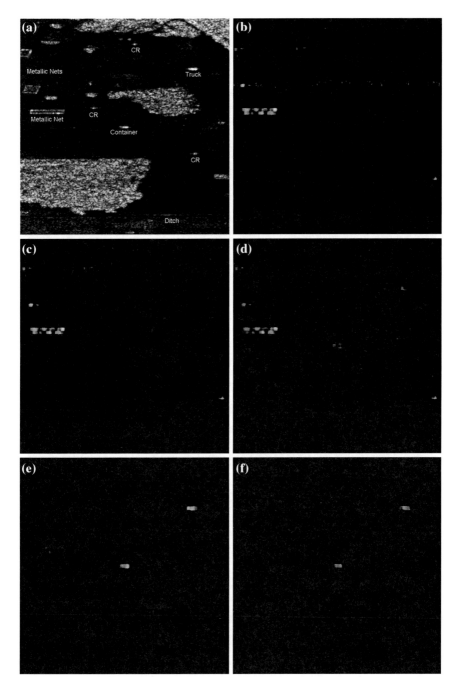

Fig. 6.12 Selectable detection over open field area. $\gamma \in [0, \pi/2]$, $\vartheta_m = \tau_m = 0$ and $\upsilon = \pi/2$. **a** L-band RGB Pauli image; **b** $\gamma = 0$; **c** $\gamma = \pi/8$; **d** $\gamma = \pi/4$; **e** $\gamma = 3\pi/8$; **f** $\gamma = \pi/2$

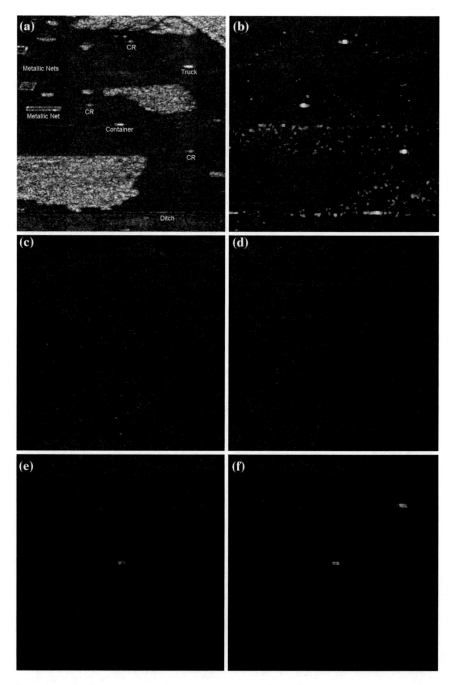

Fig. 6.13 Selectable detection over open field area. $\upsilon \in [0, \pi/2]$, $\vartheta_m = \tau_m = 0$ and $\gamma = \pi/2$. **a** L-band RGB Pauli image; **b** $\upsilon = 0$; **c** $\upsilon = \pi/8$; **d** $\upsilon = \pi/4$; **e** $\upsilon = 3\pi/8$; **f** $\upsilon = \pi/2$

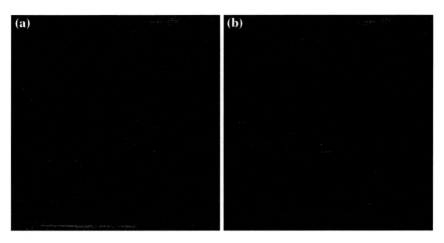

Fig. 6.14 Detection over entire test area. **a** RGB mask for multiple reflection detector. *Red*: Even-bounce, *Green*: zero, *Blue*: Odd-bounce (5 × 5); **b** RGB mask for oriented dipole detector. *Red*: horizontal dipole, *Green*: zero, *Blue*: vertical dipole (5 × 5). The intensity of the masks is related to the detector amplitude. Threshold: 0.97

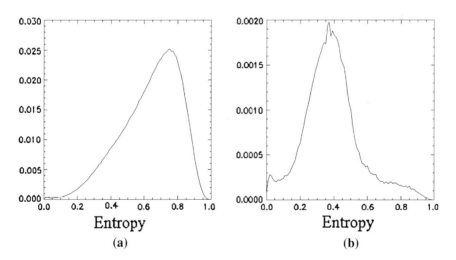

Fig. 6.15 Normalised histogram of the entropy for **a** total image and **b** detected mask

Several other experiments are possible, which can exploit different representations as well, but for the sake of brevity we will only present these four examples. In conclusion, in the case of natural targets the sensitivity analysis can help the adjustment process for the target to be detected.

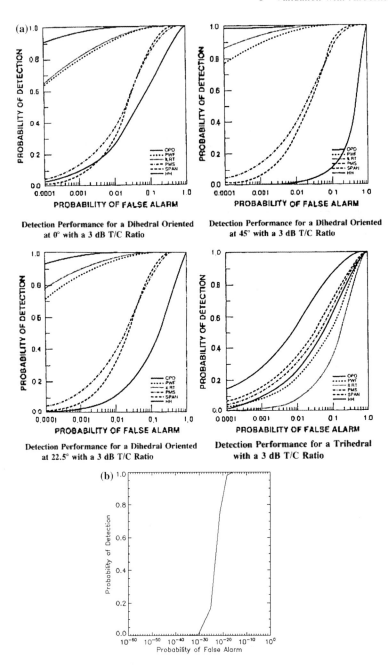

Fig. 6.16 ROC comparison among several detector. **a** *OPD*: optimal polarimetric detector, *PWF*: polarimetric whitening filter, *ILRT*: identity likelihood-ratio-test, *PMS*: power maximisation synthesis (Chaney et al. 1990). **b** Proposed detector: *SCR* = 2 and averaging window 5 × 5. The probability of false alarm is expressed in dB

6.5 Polarimetric Characterisation

This section is dedicated to the investigation of the polarimetric properties of the detected targets. The detector is based on the concept of scattering vector posing an implicit restriction to single targets.

Here, we intend to test the degree of polarisation of the detected points, confirming their nature as single targets.

In the third chapter the entropy was introduced as an estimator of the degree of polarisation of the target, or in other words a measure of how close the target is to being a single scatterer. A low entropy signifies presence of a single dominant scattering mechanism (Cloude and Pottier 1997).

Figure 6.14 shows the detection masks over the entire test area: 1400×1400 pixels (the RGB of the area is presented in Fig. 6.2). A noteworthy aspect of the detector, not specified previously is its speed of processing. To produce four detection masks on the entire test area the detector took less than 5 s using IDL on a standard desktop with 2 GHz RAM (please note the time to load the images as IDL variables is not considered). This suggests that the detector could be employed in real time.

In order to assess the degree of polarisation, the entropy for all the detected points on the mask is estimated. Figure 6.15 shows the histogram of the entropy of the detected points (Fig. 6.15b) compared with the entropy of all the points in the scene (Fig. 6.15a). The former have entropy generally lower than 0.5, indicating targets with rather single behaviour.

6.6 Comparison with Another Polarimetric Detector: PWF

The issue investigated in this section is the comparison of the developed detector with pre-existent polarimetric detectors. In the fifth chapter, the algorithm's theoretical performance was calculated based on the statistical characterisation of the detector (its *pdf*). In particular, the *ROC* was estimated (Kay 1998). Additionally, in the third chapter the *ROC* of a few detectors was presented (Chaney et al. 1990). In order to have an easier comparison of the graphs both the figures are shown again in Fig. 6.16.

The theoretical performance of the proposed detector is several orders of magnitude superior to any of the examined detectors (especially if it is considered that we do not use statistical a priori information).

We redirect the reader to Chap. 5 for an exhaustive explanation of these results. Briefly, this is related to the reduction of variability of the noisier clutter components performed by the selected basis of the polarimetric space and the averaging.

In Fig. 6.16a, the Polarimetric Whitening Filter PWF (Novak et al. 1993; Chaney et al. 1990) seems to have the best performance among detectors without statistical a priori information about target and clutter. Moreover, it was

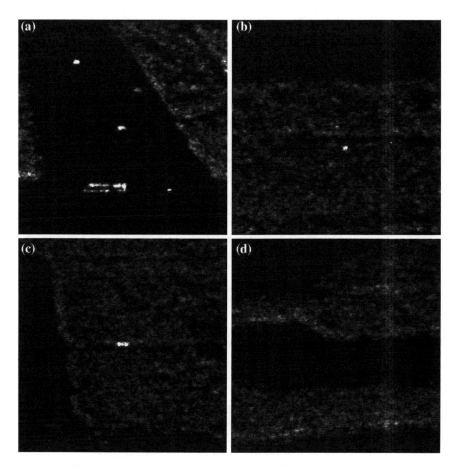

Fig. 6.17 Polarimetric whitening filter: **a** Wolf1 in open field. **b** 3 corner reflector in forested area; **c** container in forested area. **d** agricultural field (vegetated area: small stokes)

demonstrated to be the optimum solution for speckle reduction and target detection (Novak and Hesse 1993). Therefore, it appears to be the best candidate for a comparison with the proposed detector. Briefly, the PWF uses the polarisation to filter the images, and thus reduces (optimally) the speckle. Consequently, all the pixels interpreted as affected by speckle are strongly reduced, while the coherent ones are magnified.

Figure 6.17 shows the results of the PWF for four areas already presented during the analysis of standard targets (from Figs. 6.5, 6.6, 6.7, 6.8).

In open field, (Fig. 6.17a) the performance is comparable (as long as the backscattering from the target is strong enough). Both the techniques detect jeep, net and corner reflectors. However, PWF is not able to classify the detected targets, since it uses the polarimetric information to reduce the speckle. For instance, it is not possible to separate the jeep from the net or the corner reflectors.

In a more challenging environment such as detection of targets beneath forest canopy (Fig. 6.17b, c) (Fleischman et al. 1996), the PWF fails in detecting one *CR* (bottom left: 70 cm). In fact, an embedded target can present some speckle variation due to the surrounding clutter which is strong and speckled. On the other hand, the detector proposed in this thesis is not an algorithm for speckle reduction and can detect speckled targets, as long as the polarimetric behaviour is still single.

Regarding targets with weak backscattering (Fig. 6.17d), PWF is based on a threshold over the power of the detection image (which is a speckle-reduced replica), hence weak targets are completely lost (e.g. bare soil, non metallic targets). Conversely, the proposed detector is based on the weight of the target components, hence it can detect low backscattering targets, as long as they are polarimetricaly dominant. This is particularly evident in the agricultural area illustrated in Fig. 6.17d where the stripe-like field with small vertical stokes disappears completely in the processed image. The same happened to the weak targets in all the other detection images.

References

Attema EPW, Ulaby FT (1978) Vegetation modeled as water cloud. Radio Sci 13:357–364

Cameron WL (1996) Simulated polarimetric signatures of primitive geometrical shapes. IEEE Trans Geosci Remote Sensing 34:793–803

Campbel JB (2007) Introduction to remote sensing. The Guilford Press, New York

Chaney RD, Bud MC, Novak LM (1990) On the performance of polarimetric target detection algorithms. IEEE Aerospace Electron Syst Mag 5:10–15

Cloude SR (1987) Polarimetry: the characterisation of polarisation effects in EM scattering. Electronics Engineering Department, University of York, York

Cloude RS (1995a) An introduction to wave propagation & antennas. UCL Press, London

Cloude SR (1995b) Lie groups in EM wave propagation and scattering. In: Baum C, Kritikos HN (eds) Electromagnetic symmetry. Taylor and Francis, Washington, pp 91–142. ISBN 1-56032-321-3 (chapter 2)

Cloude SR (2009) Polarisation: applications in remote sensing. Oxford University Press, Oxford. ISBN 978-0-19-956973-1

Cloude SR, Pottier E (1997) An entropy based classification scheme for land applications of polarimetric SAR. IEEE Trans Geosci Remote Sensing 35:68–78

Cloude SR, Corr DG, Williams ML (2004) Target detection beneath foliage using polarimetric synthetic aperture radar interferometry. Waves Random Complex Media 14:393–414

Curlander JC, Mc donough RN (1991) Synthetic aperture radar: systems and signal processing. Wiley, New York

Dong Y, Forster B (1996) Understanding of partial polarization in polarimetric SAR data. Int J Remote Sens 17:2467–2475

Durden SL, Van Zyl JJ, Zebker HA (1999) Modeling and observations of the radar polarization signatures of forested areas. IEEE Trans Geosci Remote Sensing 27:2363–2373

Fleischman JG, Ayasli S, Adams EM (1996) Foliage attenuation and backscatter analysis of SAR imagery. IEEE Trans Aerospace Electron Syst 32:135–144

Fung AK, Ulaby FT (1978) A scatter model for leafy vegetation. IEEE Trans Geosci Electron 16:281–286

Hajnsek I, Schön H, Jagdhuber T, Papathanassiou K (2007) Potentials and limitations of estimating soil moisture under vegetation cover by means of PolSAR. In: Fifth international symposium on retrieval of bio and geophysical parameters from SAR data for land applications, Bari, September 2007

Horn R, Nannini M, Keller M (2006) SARTOM Airborne campaign 2006: data acquisition report. DLR-HR-SARTOM-TR-001

Huynen JR (1970) Phenomenological theory of radar targets. Delft Technical University, Delft

Kay SM (1998) Fundamentals of statistical signal processing. Volume 2: detection theory. Prentice Hall, Englewood Cliffs

Lang RH (1981) Electromagnetic scattering from a sparse distribution of lossy dielectric scatterers. Radio Sci 16:15–30

Lee JS, Pottier E (2009) Polarimetric radar imaging: from basics to applications. CRC Press, Boca Raton

Li J, Zelnio EG (1996) Target detection with synthetic aperture radar. IEEE Trans Aerospace Electron Syst 32:613–627

Mott H (2007) Remote sensing with polarimetric radar. Wiley, Hoboken

Novak LM, Hesse SR (1993) Optimal polarizations for radar detection and recognition of targets in clutter. In: Proceedings of IEEE national radar conference, Lynnfield, MA, pp 79–83

Novak LM, Burl MC, Irving MW (1993) Optimal polarimetric processing for enhanced target detection. IEEE Trans Aerospace Electron Syst 20:234–244

Novak LM, Owirka GJ, Weaver AL (1999) Automatic target recognition using enhanced resolution SAR data. IEEE Trans Aerospace Electron Syst 35:157–175

Papathanassiou KP, Cloude SR (2001) Single-baseline polarimetric SAR interferometry. IEEE Trans Geosci Remote Sensing 39:2352–2363

Rose HE (2002) Linear algebra: a pure mathematical approach. Birkhauser, Berlin

Rothwell EJ, Cloud MJ (2001) Electromagnetics. CRC Press, Boca Raton

Strang G (1988) Linear algebra and its applications, 3rd edn. Thomson Learning, USA

Stratton JA (1941) Electromagnetic theory. McGraw-Hill, New York

Treuhaft RN, Cloude RS (1999) The structure of oriented vegetation from polarimetric interferometry. IEEE Trans Geosci Remote Sensing 37:2620–2624

Treuhaft RN, Siqueria P (2000) Vertical structure of vegetated land surfaces from interferometric and polarimetric radar. Radio Sci 35:141–177

Tsang L, Kong JA, Shin RT (1985) Theory of microwave remote sensing. Wiley, New York

Ulaby FT, Elachi C (1990) Radar polarimetry for geo-science applications. Artech House, Norwood

Woodhouse IH (2006) Introduction to microwave remote sensing. CRC Press, Taylor and Francis Group, Boca Raton

Zebker HA, Van Zyl JJ (1991) Imaging radar polarimetry: a review. In: Proceedings of the IEEE, p 79

Chapter 7
Validation with Satellite Data

7.1 Introduction

The previous chapter was focused on validation with airborne data, while the current is specifically assigned to satellite data. This separation was beneficial due to some differences between airborne and satellite data. Although the data processing is substantially the same, there are significant practical differences making satellite detection more challenging (Campbel 2007; Chuvieco and Huete 2009).

One of the main disadvantages is represented by the rather course resolution. The resolution (as explained in the second chapter) is apparently independent of the distance between antenna and scene, while it is conditional dependent upon the available bandwidth (i.e. the range resolution is linked to the chirp bandwidth) and the amount of acquirable data per scene (i.e. the dimension of the synthetic antenna). Unfortunately, for space applications the bandwidth is limited and there are limitations on the amount of data that can be stored and sent down, coarsening the resolution in both the dimensions (Chuvieco and Huete 2009).

Another drawback is the lower Signal to Noise Ratio (SNR) achievable, since the received power goes down with the forth power of the distance since it propagates as a spherical wave in a two way trip. Therefore, the amount of energy received is lower for a high altitude platform. Additionally, the amount of energy which can be transmitted in the single pulse is limited due to an economic use of the satellite batteries (Chuvieco and Huete 2009).

On the other hand, satellite acquisitions have remarkable advantages compared to an airborne sensor, so that applications that are able to exploit satellite data are generally favourable. Satellites are always available (except for short periods of maintenance) performing several and periodic passes over the same scene. Airborne systems generally cannot offer the same attainability because a new campaign must be organised every time an acquisition is required. Additionally, the coverage of a satellite image is generally wider than an airborne system since the footprint (in the range direction) is much larger. In some applications which

A. Marino, *A New Target Detector Based on Geometrical Perturbation Filters for Polarimetric Synthetic Aperture Radar (POL-SAR)*, Springer Theses, DOI: 10.1007/978-3-642-27163-2_7, © Springer-Verlag Berlin Heidelberg 2012

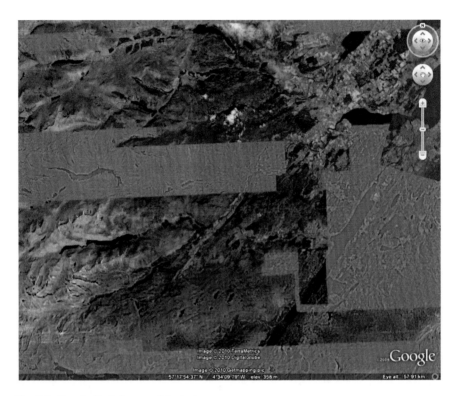

Fig. 7.1 Google Earth image of the test area

require the monitoring of waste areas (i.e. ocean surveillance) a larger footprint was revealed to be a winning point (Campbel 2007).

Although, satellite data have relevant practical advantages, they represent a more challenging scenario for detection algorithms. This is the reason behind the preference of an initial validation with airborne data: using an easier scenario the real potentials of the detector were demonstrated. In this chapter, we want to test the feasibility of the algorithm for satellite data, nonetheless we expect that the detection of small targets will be challenging. An exception is represented by TerraSAR-X which provides higher resolution data than usually available. Unfortunately, at the moment of the compilation of this thesis, quad polarimetric TerraSAR-X data were still experimental and only one scene was distributed to the public.

The proposed polarimetric detector operates on the full geometric space of single targets which can be thoroughly reconstructed only with quad polarimetric data (Cloude 2009; Huynen 1970; Kennaugh and Sloan 1952; Lee and Pottier 2009; Mott 2007; Deschamps and Edward 1973; Ulaby and Elachi 1990; Zebker and Van Zyl 1991). Therefore, the validation can be performed only on satellite systems able to acquire this class of data. Nowadays, three such satellites are available: ALOS PALSAR (L-band) (ALOS 2007), RADARSAT2 (C-band) (Slade 2009) and TerraSAR-X (X-band) (Fritz and Eineder 2009). We decided to

Fig. 7.2 RGB Pauli
composite image of the entire
dataset. *Red HH − VV*, *Green*
2HV, *Blue HH + VV*

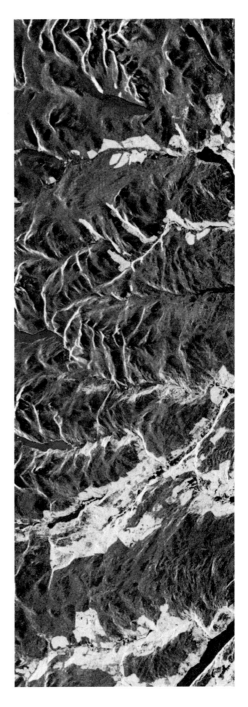

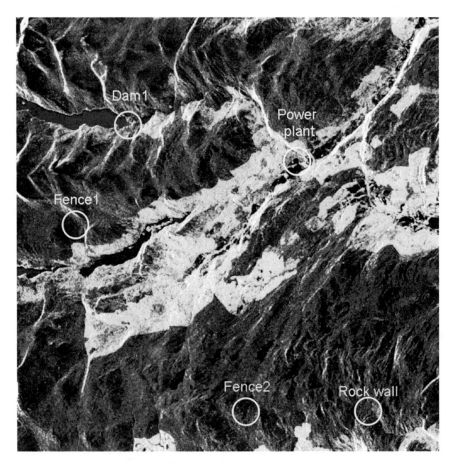

Fig. 7.3 RGB Pauli composite image of Loch Benin. *Red HH − VV, Green 2HV, Blue HH + VV*

test the detector with all of them. Together, they depict a rather fascinating scenario with different central frequencies and resolutions. For this reason, it is expected that they will return a reasonably broad picture of the detection capabilities.

7.2 ALOS PALSAR

7.2.1 Description of the Data and General Considerations

ALOS-PALSAR is the quad polarimetric radar system of the Japanese Space Exploration Agency (JAXA) (ALOS 2007). The carrier frequency of the pulse is in L-band (1.270 GHz), with a corresponding wavelength of about 23 cm. It shares

Fig. 7.4 Google Earth image of area around Loch Benin

the same band as the E-SAR system, subsequently the physical targets visible on the scene should be similar to those observed in the previous chapter. On the other hand, the resolution of ALOS is much lower than E-SAR (with 14 MHz bandwidth and about 4.5 × 30 m in azimuth and ground range). A coarse resolution makes the detection of small targets rather challenging, since their return is spread over a larger area with the possibility of being submerged by the surrounding clutter. For this reason, we do not expect to detect small vehicles, posts, huts, etc. (as in the previous chapter) (Li and Zelnio 1996; Novak et al. 1999).

The dataset used to test the detector was acquired in the area of Glen Affric (Orthographic 57.256; −5.019) in Scotland on the 18th of April 2007. This is a relatively uninhabited region, with a few sparse constructions (generally small). Conversely, the area is remarkable from an ecological point of view since there is an old Caledonian Pine forest (one of the few left in Scotland) (Forestry-Commission 2010).

Figure 7.2 shows the RGB Pauli composite image of the entire dataset (Cloude 2009; Lee and Pottier 2009). Again red represents $HH - VV$ (even bounce or even number of reflections), blue is $HH + VV$ (odd bounce or odd number of reflections) and green is twice HV (45° oriented even bounce). The direction of flight is vertical from bottom to top (i.e. ascendant orbit). The image presented here was multi-

Fig. 7.5 Detection of multiple reflections over Loch Beneveian. *Red* even-bounce, *Green* zero, *Blue* odd-bounce; The intensity of the masks is related to the detector amplitude

looked in the azimuth direction 5 times with the intention of making the pixel approximately squared on the ground. As mentioned previously, the ALOS resolution cell is not square with the azimuth about 5 times smaller than the ground range (ALOS 2007). This leads to a severe distortion when the radar image is compared to a map making the reflectivity image hard to interpret (Franceschetti and Lanari 1999).

The multi-look process needs clarification. In order to preserve the polarimetric information the multi-look cannot be performed on the scattering matrix since the relative phases would be lost, modifying the final polarimetric target. In this thesis, the multi-look was performed on the coherence [*T*] or covariance [*C*] matrices, where all the matrix elements were multi-looked separately (Lee et al. 1993, 1994).

Figure 7.1 illustrates the aerial photograph of the region (Google Earth) with the intention of facilitating the interpretation of the features visible in the RGB image.

Fig. 7.6 Detection of oriented dipoles over Loch Beneveian. *Red* horizontal dipole, *Green* zero, *Blue* vertical dipole; The intensity of the masks is related to the detector amplitude

The image is composed by a mosaic of aerial photographs with different resolutions and only a small section of the image has high resolution. Although, the multi-look makes the radar resolution cell almost squared, we still cannot overlap the radar and the optical image without a geo-coding stage (Woodhouse 2006).

In Fig. 7.2, the lochs are features with a characteristic polarimetric response which makes them easily separable from the rest of the scene. In the RGB image they appear blue since the water surface can be modelled with a Bragg surface and it is rather similar to the ideal surface represented by *HH* + *VV* (or odd-bounce).

The rest of the scene is generally bluish as a consequence of the surface scattering over the hills. Note that they are not as blue as the lochs because the hills are generally covered with a short layer of vegetation (mainly grass and bushes) which marginally perturbs the surface scattering introducing a small component of volume scattering (Ulaby and Elachi 1990; Zebker and Van Zyl 1991). Even

Fig. 7.7 Google Earth image of fence1 around Loch Benin

though the RGB image was obtained with a supplementary 3×3 average (over the multi-looked image) the overall image seems nosier than the E-SAR data. For instance, the red stripes are processing artefacts and they are visible especially where the signal is low (e.g. the lochs).

7.2.2 *Standard Targets Detection*

All the detections performed in this chapter are focused on standard targets: odd-bounce, even-bounce, horizontal and vertical dipoles. As explained more in detail in the previous chapter, these are selected for their relative abundance in a radar image. The section dedicated to sensitivity analysis starting from a parameterisation is skipped in this chapter. It was an exercise of tuning over the actual target to detect, however now we do not have accurate ground truth and the interpretation of the detected targets is more challenging with course resolution.

The detections are performed on portions of the total image in order to provide a closer look at the targets in the scene. However, we want to stress that the

Fig. 7.8 Google Earth image of fence2 around Loch Benin

detection algorithm is particularly fast performing the scan of the entire scene in few seconds.

Figure 7.3 shows the RGB Pauli composite image of the first portion analysed. This section represents the Southern part of the total image and has an extension of 1,248 × 1,248 pixels (the total number of pixels in range is 1,248). On the ground it covers approximately 30 km per side.

In order to locate the scene geographically, Fig. 7.4 presents the aerial photograph (Google Earth), where the loch in the middle left is Loch Benevean. Additionally, we provided the aerial photograph with labels marking some of the detected targets.

The resulting mask for multiple reflections is illustrated in Fig. 7.5. The colour coding used by the mask is the same presented in the previous chapter (red: even-bounce; blue: odd-bounce). As expected, the lochs are detected as surfaces as well as several other areas often related to ground in layover (e.g. crests of the hills). The normal angle of the surface in layover can be close to the look angle offering a more ideal odd bounce return (i.e. mirror).

As explained in the pervious chapter, even bounces are targets characterised by an even number of reflection suffered by the radiation before this can reach the antenna. In a SAR scene, the main targets responsible for even bounces are

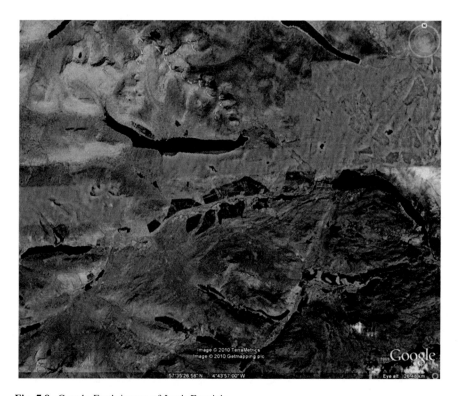

Fig. 7.9 Google Earth image of Loch Fannich

dihedrals (i.e. double bounce). With high resolution data, vehicles, houses and trunks can easily form dihedrals, on the other hand, with courser resolution the latter are not visible and only large dihedral features can be observed as even bounces.

In Fig. 7.5 the even bounces detected appear in red, and are marked in the RGB image with circles. The point close to the upper left corner of the image is a dam in Loch Mullardoch, while the one in the middle right is a power plant. Finally, the point on the bottom right corner is a rock wall. Unfortunately, the aerial photographs of the area have poor resolution and we could not find a counterpart to all the detected points.

The oriented dipoles detection is displayed in Fig. 7.6, where the horizontal dipoles seem to outnumber the vertical ones. Two of the detected points are again the dam and the power plant, nevertheless the mask shows several other spots distributed all around the hills. As mentioned previously, this region is relatively inhabited, therefore it is intriguing to identify the sources of this polarimetric return. In the RGB image (Fig. 7.3) these points appear as purple (mixture of red and blue), revealing the actual presence of scatterers different from the surroundings. Therefore, they do not seem to be false alarms due to the detector processing. A closer look at the aerial photographs of the mosaic where the

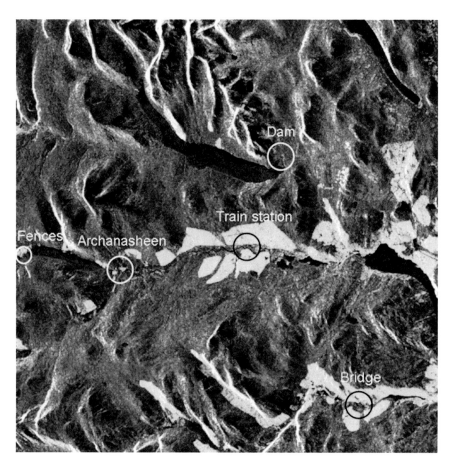

Fig. 7.10 RGB Pauli composite image of Loch Fannich. *Red HH − VV, Green 2HV, Blue HH + VV*

resolution is higher provides their identification. Those points are fences composed of horizontal wires and oriented along the flight direction (as previously observed in the E-SAR dataset). Figures 7.7 and 7.8 illustrate the high resolution aerial photographs of two areas where the detector identified horizontal dipoles. The two fences are visible as white thin stripes oriented from top to bottom, hence parallel to the flight direction. These fences are used to separate field parcels all around the area therefore they can be the cause of detections also where the resolution of the aerial photograph is to low too test their presence.

If the detection with ALOS data is compared with the one performed with E-SAR data, the amount of detected points will appear significantly reduced (especially considering that the area covered by ALOS is wider).

The cause of the fewer detections are mainly twofold: firstly, the region is less populated, with little presence of artificial targets (i.e. buildings), and secondly, the

Fig. 7.11 Detection of multiple reflections around Loch Fannich. *Red* even-bounce, *Green* zero, *Blue* odd-bounce

resolution is coarser restricting the detection only to sufficiently big targets. Exceptions are the odd bounces since they can be detected on extended (or distributed) targets like surfaces (Cloude 2009). The resolution cell after multilooking is around 30 m per side and the detector algorithm requires a further averaging through a 3 × 3 moving window (to estimate the polarimetric coherence). Therefore, objects smaller than several tens of meters are hardly detected. Even single buildings can be easily missed out if their orientation is not fortunate in generating strong double bounces.

The aerial photograph (Google) of the second test area is presented in Fig. 7.9, while Fig. 7.10 depicts the RGB Pauli. The whole portion is located in the Northern region of the total dataset. The loch in the middle upper part is Loch Fannich, the one in the middle right is Loch Luichart and the one at the upper right corner is Loch Glascarnoch.

Fig. 7.12 Detection of oriented dipoles around Loch Fannich. *Red* horizontal dipole, *Green* zero, *Blue* vertical dipole

The scene is a mix of lochs and hills, hence we expect to detect features similar to the previous experiment. The detection for multiple reflections and oriented dipoles are displayed respectively in Figs. 7.11 and 7.12.

Lochs as well as areas in layover are again detected as surfaces. Unfortunately, the aerial photographs available on Google lacks sufficient resolution to show most of the targets. However, we spotted some of the targets from a map and searched for photographs of the area. The even bounce detected in the middle of the image overlaps with the Achanalt train station (Fig. 7.13).

This is the agglomerate of just a few buildings but they seem to be properly oriented and the dihedral had a strong return in the backward direction. The same station appears as horizontal dipole as well perhaps due to the presence of fences around the buildings. The bright horizontal dipole close to Loch Fannich is the Fannich dam (also just visible in the Google image).

Fig. 7.13 Photograph of Achanalt train station (Google Earth: Panoramio)

Fig. 7.14 Photograph of Archanasheen train station (Google Earth: Panoramio)

The cluster of horizontal dipoles detected close to Loch A Chroisg (small loch in the middle right of the image) is the village, train and petrol station of Archanasheen. Figure 7.14 shows the panoramic photo of the Archanasheen train station.

Regarding the horizontal dipole detected at the left of Loch A Chroisg, we found a picture of the valley (Fig. 7.15) which reveals the presence of several fences which could cause the horizontal dipole return as observed in the E-SAR data. Other detections are visible close the river Dalnacreich (bottom right of the image). The point on the right hand side was identified as a bridge for a walking path as illustrated in Fig. 7.16. We also believe that the other two points are related to structures around the small river.

Fig. 7.15 Photograph of Loch A Chroisg valley (Google Earth: Panoramio)

Fig. 7.16 Photograph of bridge on river Dalnacreich (Google Earth: Panoramio)

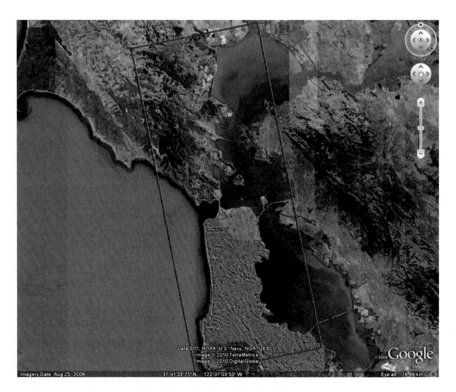

Fig. 7.17 Aerial photograph (Google Earth) of the RADARSAT-2 scene over San Francisco. The polygon shows the location of the radar scene

7.3 RADRSAT-2

7.3.1 Description of the Data

RADARSAT2 was launched by the Canadian Space Agency in December 14, 2007 and exploits a different frequency from ALOS and E-SAR, the C-band (around 5 Ghz or 5 cm wavelength) (Slade 2009). When an object is illuminated by incident radiation with a different frequency, it generally modifies its scattering behaviour (Stratton 1941; Rothwell and Cloud 2001; Cloude 1995). For this reason, the use of a different band could reveal new typologies of targets. However, the backscattering from some class of targets, as ideal reflections and dipoles is relatively independent of the frequency of the incident radiation since after the scaling performed by the change of frequency, surfaces and wires remain the same. If real rather than ideal targets are examined, the change in frequency can modify the polarimetric behaviour. However, if the frequency does not suffer a dramatic variation, the major difference for real reflections is related to the amount of backscatter more than the polarimetric characterisation, since at higher frequency

Fig. 7.18 RGB Pauli composite image of San Francisco. *Red HH — VV, Green 2HV, Blue HH + VV*

the surfaces look bigger (and collect more energy). On the other hand, narrow dipoles look thicker with possible changes in polarimetric behaviour, since they start to resemble surfaces (Cloude 2009; Rothwell and Cloud 2001). Clearly, when

Fig. 7.19 RGB Pauli composite image of San Francisco sub-area. *Red HH − VV, Green 2HV, Blue HH + VV*

the frequency variation is drastic, the target can transform completely. For instance, the reflector planes can cease to be surfaces and a narrow cylinder can become a surface.

In conclusion, from the mathematical point of view, the algorithm is apparently independence on the frequency, since the scattering vector formalism can be applied for any frequency (as long as the phase measurements are feasible). However, the scattering vector representation of the same real target can be dependent on the frequency. This is the main motivation for testing the detector with different frequencies.

The ground resolution of the single look complex (SLC) image is about 5 m in azimuth and around 10 m in range (i.e. better resolution than ALOS)(Slade, 2009). This should increase the amount of targets the algorithm is able to detect.

Fig. 7.20 Detection of multiple reflections over San Francisco. *Red* even-bounce, *Green* zero, *Blue* odd-bounce

The dataset employed represents a freely available scene of San Francisco acquired on the nineth of April 2008. Figure 7.17 illustrates an aerial photograph (Google) with a polygon showing the location of the scene.

Figure 7.18 shows the RGB Pauli composite image of the entire scene. In order to obtain a pixel approximately square on the ground, the covariance matrix was multi-looked two times along the azimuth. Again the colour coding is the same used previously. It is interesting to note that the sea is clearly identifiable in blue (i.e. interpreted as a rough surface), while the urban areas are purple (i.e. red plus blue) since the dominant scattering mechanisms in such environments are multiple reflections.

Fig. 7.21 Detection of oriented dipoles over San Francisco. *Red* horizontal dipole, *Green* zero, *Blue* vertical dipole

7.3.2 Standard Targets Detection

Again the detection was performed on a sub-area in order to provide a closer inspection of the scene. Figure 7.19 shows the RGB Pauli of the region located in the central and most urbanised area of San Francisco, between Bay Bridge and Golden Gate Bridge. The image is 1,500 × 1,500 pixels covering an area approximately of 15 × 15 km. Blocks of houses and quarters are identifiable as reddish areas (in general) with a peculiar texture. On the other hand, the sea is clearly recognisable in blue and it can be easily discriminated from the land region. Some ships can be observed as bright spots in the lower right corner of the RGB image. In particular, the Golden Gate Bridge exhibits a curious scattering behaviour, where three different returns can be identified (Woodhouse 2006). The flight track of the satellite runs

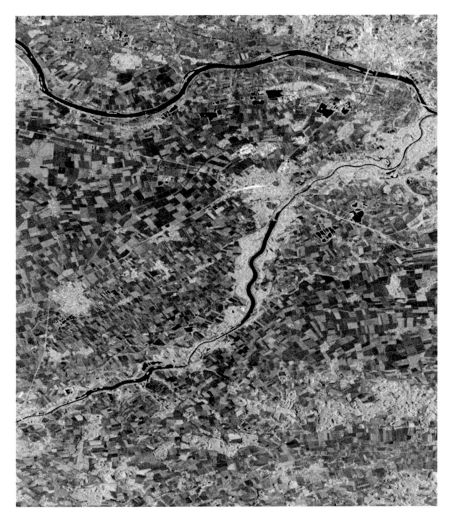

Fig. 7.22 RGB Pauli composite image of Deggendorf, Germany. *Red HH − VV, Green 2HV, Blue HH + VV*

bottom to top (i.e. ascending orbit) and the platform was right-looking, consequently the return on the left hand side is the closest to the sensor (the range axis grows from left to right). This is due to the direct scattering from the bridge structure which is in layover with the ocean surface about 200 m below.

The second return (the one in the middle) is due to double bounces with the ocean surface. The extra path travelled by the wave which reflects on the two dihedral planes always sum up into the distance of the dihedral corner (this is the reason why all the contributions sum coherently concentrating the energy of the incident radiation). The third return is a consequence of triple bounces (or more generally triple interactions) among the bridge structures and ocean which creates

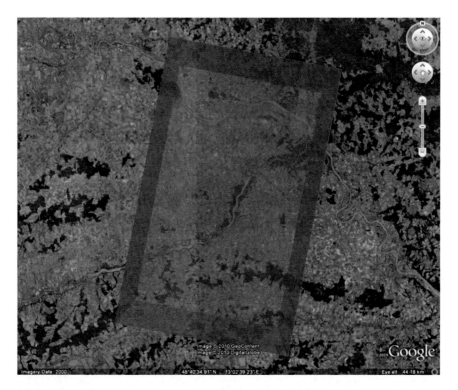

Fig. 7.23 Google Earth image of the TerraSAR-X scene on Deggendorf, Germany. The polygon shows the location of the scene

an extra path locating the return after the real position of the bridge (Woodhouse 2006). Regarding the Bay Bridge (on the right hand side), it does not present the same particular scattering behaviour because its orientation is less favourable to generate dihedrals focused along the range direction (therefore only two scattering mechanism are generated, direct and general triple interactions).

The detection masks are presented in Figs. 7.20 and 7.21. As expected, the water body is largely detected as single bounce since it can be modelled as a Bragg surface. On the other hand, most of the building blocks are detected as double bounces, due to the dihedral formed between concrete walls and tarmac. The second return from the bridge is detected as even-bounce between bridge and ocean surface confirming our interpretation. Additionally, some portion of the first return is identified as even-bounces, due to even reflections among close structures of the bridge (and not the ocean). The third return from the bridge is not identified as odd-bounces because in the interaction "water-bridge-water" the bridge generally does not present surfaces with the right orientation to close the trihedral plates and a general scattering takes place (as for the first return). Three ships are also detected in the ocean area. Clearly, in a radar image not every ship has a strong double bounce with the sea since this scattering mechanism is strongly

Fig. 7.24 RGB Pauli composite image of Doggendorf and river Donau. *Red HH − VV, Green 2HV, Blue HH + VV*

related to the orientation with respect to the flight direction as well as the near instantaneous angle between the ship and the ocean. Hence, in some situations the even-bounce does not represent the best ideal target to detect ships. Recently, an alternative approach for ship detection was developed by the author, which is based on detecting any feature on the sea surface which has a polarimetric response different from the sea (Marino et al. 2010).

Finally, the mask in Fig. 7.20 presents a variant respect all the other masks, since the green colour is added to represent a new typology of targets: dihedrals with the corner 45° oriented with respect to the Line of Sight (LOS). One block of buildings in the city conglomerate reveals their presence. In the aerial photograph, that city quarter appears to have a different orientation (about 45° with respect to the flight direction), and the block is located on a steep slope. Hence, the inclination of the buildings plus the tilted ground surface produces a target similar to a 45° dihedral.

Fig. 7.25 Detection of multiple reflections of Doggendorf and river Donau. *Red* even-bounce, *Green* zero, *Blue* odd-bounce

The latter target is not ideal (giving a not perfect detection) since the steepness of the road is not 45°, but it is close enough to generate detections when the exploited threshold is lower.

Regarding the oriented dipoles, they are detected in most of the urban areas, perhaps due to wires and railings. In order to achieve a detection, the dimension of the target generally must be equal or bigger than the resolution cell (i.e. 10 m).

A horizontal wire can easily be tens of meters long while the vertical ones are much smaller for clear reasons. More specifically, wires running along the ground range direction can be as long as horizontal one, however their return is much weaker due to their directivity pattern which does not scatter significantly in the backward direction. This is the reason why the detection is densely populated with horizontal dipoles (in the urban area) while vertical ones are quite scarce.

Fig. 7.26 Detection of oriented dipoles of Doggendorf and river Donau. *Red* horizontal dipole, *Green* zero, *Blue* vertical dipole

7.4 TerraSAR-X

7.4.1 Description of the Data

The last dataset exploited for testing the detector is a quad polarimetric TerraSAR-X dataset (Fritz and Eineder 2009). The latter was launched on June 15th, 2007 by the German Aerospace Centre DLR. It is a particularly relevant instrument for two main reasons. Firstly, the system is based on X-band, about 9.65 GHz or 3.1 cm wavelength. Again, a different frequency can explore different typologies of targets, although the standard targets investigated should not change drastically (except for a backscattering factor). Secondly, for its higher resolution (1 m in slant range and 6 m in azimuth) compared with the other two satellite systems.

Fig. 7.27 RGB Pauli composite image of Langenisarhofen. *Red HH − VV, Green 2HV, Blue HH + VV*

As consequence of the higher resolution the detection with TerraSAR-X is expected to improve with respect to ALOS and RADARSAT-2.

Unfortunately, the proposed detector requires quad polarimetric data to reconstruct the full polarimetric space. This mode in TerraSAR-X is only experimental and at the time of the compilation of this thesis only one quad polarimetric acquisition was made available to the scientific community. Although, with the launch of Tandem-X (a twin of TerraSAR-X exploited to collect the DSM of the world) the quad polarimetric mode should be more available.

Figure 7.22 shows the RGB image of the entire dataset after a multi-look of 5 × 5, which is actually the window used by the detector. The location of the scene is particularly advantageous for researchers working with classification since it represents a mixture of several agricultural fields, forests stands, urban areas and water. Moreover the topography is particularly flat helping the classification process.

Fig. 7.28 Detection of multiple reflections of Langenisarhofen. *Red* even-bounce, *Green* zero, *Blue* odd-bounce

Most of the fields appear in a bluish colour since the rough surface is the main scattering mechanism. It is interesting to note that different fields often have different colours in the RGB image, underling a different polarimetric behaviour, dependent on the soil roughness and the amount/variety of vegetation.

The urban areas are still displayed in purple (i.e. mixture of reflections). The corresponding area is illustrated in Fig. 7.23 in an aerial photograph (Google Earth) with a polygon indicating the acquired portion. The latter is close to the conjunction of two rivers, the Donau and the Isar (i.e. the river crossing Munich). The two bigger towns in the scene are Doggendorf (upper part) and Plattling (middle part).

Fig. 7.29 Google Earth image of the TerraSAR-X scene of Moosmuhle. Electric line pillars

7.4.2 Standard Target Detection

As in the previous experiments, the detection is focused on multiple reflections and oriented dipoles and portions of the total image are analysed separately in order to provide a closer look at the targets detected in the scene.

Figure 7.24 presents the RGB Pauli of the section located in the conjunction of the Donau and Isar rivers. The town of Deggendorf can be identified in the upper part of the image. This area was selected with the intention of focusing the detector on the urban area of Doggendorf. Moreover, rivers already revealed the presence of targets of interest.

The detection of multiple reflections is presented in Fig. 7.25. Any conglomerate of buildings (town and villages) reveals detected points, confirming the effectiveness of X-band and the higher resolution. As in the previous datasets, the bridges (especially the one on the right hand side) are detected as even bounces. Clearly, an individual small building can be missed by the detector since the resolution is still greater than ten meters and dihedrals are dependent on the inclination of the wall with respect to the flight direction.

Fig. 7.30 Detection of oriented dipoles of Langenisarhofen. *Red* horizontal dipole, *Green* zero, *Blue* vertical dipole

On the other hand, in a conglomerate of buildings the copious presence of walls and corners is generally sufficient to generate even reflections focused in the backward direction.

Figure 7.26 shows the oriented dipoles mask. Again, most of the detections are located in the urban area, where horizontal wires exceed in number the vertical ones. However, with the finer resolution of TerraSAR-X, it is possible to identify some vertical dipoles as well.

Considering that most of the detections are inside the urban area, it is not necessary to bring high resolution aerial photograph to validate them (since we know they correspond mainly to buildings).

As a last experiment, the detection is performed on another area of the dataset. Figure 7.27 represents the RGB image.

The typology of targets included in the marker (ellipse) will be described in the following. The village in the middle right is Langenisarhofen, while the one on the middle left is Aholming.

Figure 7.28 shows the detection of multiple reflections. Again the urban areas reveal several detection points. In the mask there is a curious alignment of even bounce targets running with an approximate inclination of 45°across the image. A closer inspection on the high resolution aerial photograph reveals the presence of big electric pylons. In the RGB (Fig. 7.27), the line of pillars is marked with a white ellipse and after a scrupulous examination the pillars can be identified in the radar image. Figure 7.29 presents the aerial photograph of the area in the upper left part of the ellipse, revealing the pylons. Their height is appreciable from the shadows they produce on the ground.

The dipoles detection is presented in Fig. 7.30. The detections are mainly associated with buildings. In this experiment, the electric wires cannot generate horizontal dipoles due to their inclination.

References

ALOS (2007) Information on PALSAR product for ADEN users

Campbel JB (2007) Introduction to remote sensing. The Guilford Press, New York

Chuvieco E, Huete A (2009) Fundamentals of satellite remote sensing. Taylor & Francis Ltd, Boca Raton

Cloude RS (1995) An introduction to wave propagation & antennas. UCL Press, London

Cloude SR (2009) Polarisation: applications in remote sensing. Oxford University Press, New York. doi:978-0-19-956973-1

Deschamps GA, Edward P (1973) Poincare sphere representation of partially polarized fields. IEEE Trans Antennas Propag 21:474–478

Forestry-Commission (2010) http://www.forestry.gov.uk/scotland

Franceschetti G, Lanari R (1999) Synthetic aperture radar processing. CRC Press, New York

Fritz T, Eineder M (2009) TerraSAR-X, ground segment, basic product specification document. DLR, Cluster Applied Remote Sensing, Weßling

Huynen JR (1970) Phenomenological theory of radar targets. Technical University, Delft, The Netherlands

Kennaugh EM, Sloan RW (1952) Effects of type of polarization on echo characteristics. In: Ohio State University Research Foundation Columbus Antenna Lab (ed)

Lee JS, Pottier E (2009) Polarimetric radar imaging: from basics to applications. CRC Press, Taylor and Francis Group, Boca Raton

Lee JS, Hoppel KW, Mango SA, Miller AR (1993) Intensity and phase statistics of multi-look polarimetric SAR imagery. IGARSS Geoscience and remote sensing symposium, vol 32

Lee JS, Jurkevich I, Dewaele P, Wambacq P, Oosterlinck A (1994) Speckle filtering of synthetic aperture radar images: a review. Remote Sens Rev 8:313–340

Li J, Zelnio EG (1996) Target detection with synthetic aperture radar. IEEE Trans Aerosp Electron Syst 32:613–627

Marino A, Walker N, Woodhouse IH (2010) Ship detection using SAR polarimetry. The development of a new algorithm designed to exploit new satellite SAR capabilities for maritime surveillance. In: Proceedings on SEASAR 2010, Frascati, Italy, Jan 2010

Mott H (2007) Remote sensing with polarimetric radar. John Wiley & Sons Inc, Hoboken

Novak LM, Owirka GJ, Weaver AL (1999) Automatic target recognition using enhanced resolution SAR data. IEEE Trans Aerosp Electron Syst 35:157–175

Rothwell EJ, Cloud MJ (2001) Electromagnetics. CRC Press, Boca Raton

Slade B (2009) RADARSAT-2 product description. Dettwiler and Associates, MacDonals

Stratton JA (1941) Electromagnetic theory. McGraw-Hill, New York

Ulaby FT, Elachi C (1990) Radar polarimetry for geo-science applications. Artech House, Norwood

Woodhouse IH (2006) Introduction to microwave remote sensing. CRC Press, Taylor & Frencies Group, Boca Raton

Zebker HA, Van Zyl JJ (1991) Imaging radar polarimetry: a review. In: Proceedings of the IEEE, vol 79

Chapter 8
Recent Applications
of Perturbation Filters

8.1 Introduction

As explained in the abstract, the main goal of the thesis was the development of a
new algebraic procedure for target detection, based on a novel perturbation filter.
The earliest and most studied application was the single target detector (*STD*)
largely described in the previous chapters. However, the algebraic operation
underneath the *STD* is more general and powerful and can be adapted to different
scenarios (as long as the entities under analysis lie within a Euclidean space).

Very recently, work was published on a version of the algorithm able to detect
partial (i.e. depolarised) targets. The aim of this chapter was to present these most
recent outcomes. Please note that work is still in progress on this partial target
detector (*PTD*). The latest results occurred after the compilation of this thesis.
Therefore, they could not be the main focus of the entire manuscript. However, the
author believes that due to their significance they could be included, even if
briefly, in the present chapter.

As shown previously, an *STD* is a powerful tool to detect man-made structures.
However, this tool is unable to detect partial or depolarised targets. Once a *PTD* is
developed, it can be easily exploited as a supervised classifier. In this chapter, the
derivation of the new algorithm is briefly presented. Subsequently, it will be tested
on several datasets in order to reveal its performance.

8.2 Partial Target Detector

8.2.1 Formulation

As expressed in the introduction chapter on polarimetry, there is a fundamental
difference between single and partial targets. Specifically, partial targets cannot be

A. Marino, *A New Target Detector Based on Geometrical Perturbation Filters for*
Polarimetric Synthetic Aperture Radar (POL-SAR), Springer Theses,
DOI: 10.1007/978-3-642-27163-2_8, © Springer-Verlag Berlin Heidelberg 2012

completely and uniquely characterised by a single scattering matrix $[S]$. They need
the formalism called coherence matrix:

$$[C] = \langle \underline{k}\, \underline{k}^{*T} \rangle = \begin{bmatrix} \langle |k_1|^2 \rangle & \langle k_1 k_2^* \rangle & \langle k_1 k_3^* \rangle \\ \langle k_2 k_1^* \rangle & \langle |k_2|^2 \rangle & \langle k_2 k_3^* \rangle \\ \langle k_3 k_1^* \rangle & \langle k_3 k_2^* \rangle & \langle |k_3|^2 \rangle \end{bmatrix}, \tag{8.1}$$

where $\underline{k} = [k_1, k_2, k_3]^T$ is the scattering vector in any basis as presented in Chap. 3
(Cloude 2009; Lee and Pottier 2009).

In order to extend the detectability of the algorithm to partial targets, a new
formalism similar to the one used for single targets must first be introduced. To
this end, a feature partial scattering vector is defined:

$$\begin{aligned} \underline{t} &= Trace([C]\Psi) = [t_1, t_2, t_3, t_4, t_5, t_6]^T \\ &= \left[\langle |k_1|^2 \rangle, \langle |k_2|^2 \rangle, \langle |k_3|^2 \rangle, \langle k_1^* k_2 \rangle, \langle k_1^* k_3 \rangle, \langle k_2^* k_3 \rangle \right]^T. \end{aligned} \tag{8.2}$$

where Ψ is a set of 6×6 basis matrices under a Hermitian inner product. t lies in a
subspace of C^6 (it is closed for sum and scalar multiplication and includes the
zero). In particular, the first three components are real positive and the second
three complex. To have physical feasibility the last three elements must obey the
Cauchy–Schwarz (Rose 2002) inequality

$$\langle |\underline{x}| \rangle \langle |\underline{y}| \rangle \geq \left| \langle \underline{x}^{*T} \underline{y} \rangle \right| : \tag{8.3}$$

$$\sqrt{t_1 t_2} \geq |t_4|, \quad \sqrt{t_1 t_3} \geq |t_5|, \quad \sqrt{t_2 t_3} \geq |t_6|. \tag{8.4}$$

Any physically realisable $\underline{t}$ represents completely and uniquely a partial target.
In particular, the partial target to be detected and the perturbed target are regarded as

$$\begin{aligned} \hat{\underline{t}}_T &= Trace([C_T]\Psi)/\|Trace([C_T]\Psi)\|, \\ \hat{\underline{t}}_P &= Trace([C_P]\Psi)/\|Trace([C_P]\Psi)\|. \end{aligned} \tag{8.5}$$

The latter could be seen as the equivalent of the scattering mechanisms for
partial targets. Although the optimisation of the perturbation has mathematical
foundations (Marino et al. 2009, 2010; Marino and Woodhouse 2009), physical
meaning can be attributed to the process. For instance, the covariance matrix for
the target $[C_T]$ can be mapped into a Kennaugh matrix $[K_T]$ (Cloude 2009).
Subsequently, the Huynen transformations can be performed on the Kennaugh
matrix generating a slightly different target $[K_P]$(Huynen 1970). Finally, the perturbed Kennaugh matrix $[K_P]$ is mapped back into a covariance matrix $[C_P]$ (and
the vector $\hat{\underline{t}}_P$). The latter is merely an example of physical perturbation of the
partial target and any other parameterisation can be exploited.

Again, a change of basis is performed which makes the target of interest lie only in one nonzero component:

$$\hat{t}_T = [1,0,0,0,0,0]^T \text{ and } \hat{t}_P = [a,b,c,d,e,f]^T. \tag{8.6}$$

In the case the perturbation is performed without any physical model, $\hat{t}_P$ must be selected preserving the physical feasibility:

$$a, b, c \in \mathbb{R}^+,$$

$$\sqrt{ab} \geq |d|, \ \sqrt{ac} \geq |e|, \ \sqrt{bc} \geq |f|, \tag{8.7}$$

$$a^2 + b^2 + c^2 + d^2 + e^2 + f^2 = 1.$$

Additionally, by definition of perturbed target:

$$a \gg b, \ a \gg c, \ a \gg |d|, \ a \gg |e|, \ a \gg |f|. \tag{8.8}$$

The elements on the diagonal of $[A]$ are the components of the partial scattering vector t after the change of basis which makes $\hat{t}_T = [1,0,0,0,0,0]^T$. The change of basis can be achieved by multiplying by a unitary matrix, where the columns can be derived by solving a linear equation system, where the unknowns are five rotation angles and five phase angles.

A simpler way to generate $[A]$ considers a Gram–Schmidt ortho-normalisation (GS) in $\mathbb{C}^6$, where the first axis is the vector $\hat{t}_T$. The components of $[A]$ are calculated with the inner product of the basis for the observable t. If If $\underline{u}_1 = \hat{t}_T$, $\underline{u}_2$, $\underline{u}_3$, $\underline{u}_4$, $\underline{u}_5$, and $\underline{u}_6$ represent the ortho-normal basis then

$$[A] = \text{diag}\left(\hat{t}_T^{*T} \underline{t}, \underline{u}_2^{*T} \underline{t}, \underline{u}_3^{*T} \underline{t}, \underline{u}_4^{*T} \underline{t}, \underline{u}_5^{*T} \underline{t}, \underline{u}_6^{*T} \underline{t}\right). \tag{8.9}$$

The detector can be achieved with

$$\left([A]\hat{t}_T\right)^{*T}([A]\hat{t}_P) = \hat{t}_T^{*T}[A]^{*T}[A]\hat{t}_P = \hat{t}_T^{*T}[P]\hat{t}_P, \tag{8.10}$$

where

$$[P] = \text{diag}(P_1, P_2, P_3, P_4, P_5, P_6). \tag{8.11}$$

$$\gamma_d = \frac{1}{\sqrt{1 + \dfrac{b^2}{a^2}\dfrac{P_2}{P_1} + \dfrac{c^2}{a^2}\dfrac{P_3}{P_1} + \dfrac{|d|^2}{a^2}\dfrac{P_4}{P_1} + \dfrac{|e|^2}{a^2}\dfrac{P_5}{P_1} + \dfrac{|f|^2}{a^2}\dfrac{P_6}{P_1}}}. \tag{8.12}$$

The partial detector is formally similar to the single one in Eq. 4.37 (except for the number of terms), consequently all the mathematical optimisations performed for the single target detector can be adopted here (Marino et al. 2009, 2010b; Marino and Woodhouse 2009). Specifically, in absence of a priori information about the clutter, the perturbed target is chosen as

$$b = c = |d| = |e| = |f|. \tag{8.13}$$

If we define the clutter as $P_c = P_2 + P_3 + P_4 + P_5 + P_6$, the target as $P_1 = P_T$ and $RedR = (b/a)^2$ the detector becomes

$$\gamma_d = \frac{1}{\sqrt{1 + RedR \frac{P_c}{P_T}}}. \tag{8.14}$$

The detector is finalised with a threshold T on γ_d. The resulting mask is zero if the detector is under the threshold or equal to the detector if it is above the threshold. In other words:

$$\left\{ \begin{array}{ll} m(x,y) = 0 & \text{if } \gamma_d(x,y) < T \\ m(x,y) = \gamma(x,y) & \text{if } \gamma_d(x,y) \geq T \end{array} \right\}. \tag{8.15}$$

where m is the image mask, (x,y) represents the coordinate of a generic pixel. Using this typology of mask (and not a 1 or 0 binary format), we want to preserve information about the dominance of the target in the cell. This will be useful for the design of a classifier as we show in the following.

8.2.2 Physical Feasibility

In this section, clarifications about the uniqueness and the Gram–Schmidt ortho-normalisation (GS) are provided.

The former is guaranteed since, by definition, any partial target can be described by a covariance matrix $[C]$ (specifically, 9 real independent parameters). Additionally, all the independent elements of $[C]$ are unequally mapped in the feature vector $\underline{t}$. In the proposed 6 dimensional complex space, any partial target can be uniquely related to a single feature vector $\underline{t}$, independently on the target degree of polarisation: from pure (single targets) to completely unpolarised (random noise). In conclusion, there is a 1 by 1 relationship between the physically feasible t and any partial target.

Regarding the GS, generally, the resulting basis does not represent a set of physical feasible targets, except for the first axis, which is calculated starting from a physical realisable vector $\hat{t}_T$. GS generates a basis for $\mathbb{C}^6$ but not all the vectors of $\mathbb{C}^6$ are physically feasible. This does not however represent a limitation of the detector. The axes $\underline{u}_2$, $\underline{u}_3$, $\underline{u}_4$, $\underline{u}_5$ and $\underline{u}_6$, obtained with the GS ortho-normalisation, span a subspace of $\mathbb{C}^6$ which is completely orthogonal to the first axis $\hat{t}_T$ (i.e. the orthogonal complement of $\hat{t}_T$ in $\mathbb{C}^6$). This means that given a vector

$$\underline{u} = c_2 \underline{u}_2 + c_3 \underline{u}_3 + c_4 \underline{u}_4 + c_5 \underline{u}_5 + c_6 \underline{u}_6, \tag{8.16}$$

we have

$$\underline{u}_1 = \hat{t}_T \perp \underline{u}, \ \forall c_1, c_2, c_3, c_4, c_5 \in \mathbb{C} \tag{8.17}$$

The first vector of the *GS* basis $\underline{u}_1$ is always physically realisable, since it is equal to $\hat{\underline{t}}_T$ (i.e. the target to be detected). We refer to the orthogonal complement subspace of $\hat{\underline{t}}_T$ in $\mathbb{C}^6$ as Z. Clearly only a portion (i.e. subspace) of Z represents physically feasible targets. Moreover, a physically feasible target extracted from the data, will generally have a component in the Z subspace, called $\underline{z}$. The length of $\underline{z}$ is independent of the basis used to represent Z (since the length is an invariant property of the vector z) (Rose 2002; Strang 1988). Therefore, we do not require that $\underline{u}_2$, $\underline{u}_3$, $\underline{u}_4$, $\underline{u}_5$, and $\underline{u}_6$ are physically feasible vectors, as long as they represent a basis for Z.

As Eq. 8.17 shows, we are interested in P_T, while P_C represent the rest of the power. Clearly, equal results are obtained starting from Eq. 8.17 and considering $P_C = P_{tot} - P_T$, where

$$P_{tot} = \underline{t}^{*T}\underline{t} \tag{8.18}$$

is the total power of $\underline{t}$ in the original basis. The final simplified expression of the detector is

$$\gamma_d = \frac{1}{\sqrt{1 + RedR\left(\frac{P_{tot}}{P_T} - 1\right)}} = \frac{1}{\sqrt{1 + \frac{b^2}{1-5b^2}\left(\frac{\underline{t}^{*T}\underline{t}}{\left|\underline{t}^{*T}\hat{\underline{t}}_T\right|^2} - 1\right)}} \geq T \tag{8.19}$$

Summarising, the detector obtained with the projections on the *GS* basis and the one with the total power are entirely equivalent when $b = c = |d| = |e| = |f|$ (i.e. absence of a priori information about clutter).

8.2.3 Parameter Selection

The partial target detector proposed in this paper shares the same mathematical formalism of the single target detector in (Marino et al. 2009, 2010; Marino and Woodhouse 2009). As a consequence, all the mathematical optimisations can be extended to this case. For the sake of brevity, we only present the selection of threshold and *RedR*. This can be accomplished starting from a dispersion equation based on the angular distance between the observed partial target and the one of interest.

After some algebraic manipulation of Eq. 8.17 and substituting

$$SCR = \frac{P_T}{P_C} \Rightarrow \frac{P_{tot}}{P_T} - 1 = \frac{1}{SCR}, \tag{8.20}$$

we can find the dispersion expression:

$$0 \leq \frac{P_C}{P_T} = \frac{1}{SCR} \leq \frac{1}{RedR}\left(\frac{1}{T^2} - 1\right). \tag{8.21}$$

The first inequality is consequence of the fact that the power of the clutter cannot be bigger than the total power.

Equation 8.21 exhibits a relationship among Signal to Clutter Ratio (*SCR*), threshold and *RedR*. Here, the *SCR* has a slightly different interpretation compared with classical detection. In general, it represents the ratio between the power of target and clutter located in the scene. Instead, now it corresponds to a measure of the angular distance between the observed vector (i.e. target) and the one of interest. Its selection conforms to selectivity requirements of the filter and it can be related to the target properties. In general, when the target of interest is expected to be polarimetricaly stable, a higher *SCR* can be utilised, leading to a smaller false alarm rate. With polarimetricaly stable we mean that the angular distance of its t vector instances (realisations) is small (i.e. the representation of the target is stable over all the scene). However, if the target is anticipated to change slightly over the entire scene, a smaller *SCR* is to be preferred, which leads to higher probability of detection. In the following experiments, the *SCR* for detections is chosen equal to 50, since this value seems to provide the best compromise between probability of detection and false alarm. However, common values can go from 2 to 100.

Having defined the *SCR*, two unknowns remain in Eq. 8.18. Therefore, one unknown can be expressed as function of the other. Equation 8.19 presents one of the two possible solutions of Eq. 8.21 when the equality sign is substituted:

$$RedR = SCR \bigg/ \left(\frac{1}{T^2} - 1 \right). \qquad (8.22)$$

The threshold can be freely set. In the following experiments $T = 0.98$, although any other values smaller than 1 could be theoretically employed. However, a relatively high value of T entails a smaller variance of the polarimetric coherence, which increases the statistical performances of the detector.

Once selected T, the last parameter (i.e. *RedR*) can be set. In our experiments, $RedR = 1.85$.

8.2.4 Dual Polarimetric Detection

This final section is dedicated to the use of dual polarimetric data. The proposed algorithm is based on a geometrical operation which is theoretically independent of the dimensions of the space considered, as long as it is Euclidean. Consequently, it can be exported to any Euclidean vector space. The demand of quad polarimetric data is a physical requirement, since the entire scattering matrix is needed to characterise uniquely a generic depolarised target. Using dual polarimetric data, only a portion of the target space can be explored and the target behaviour in the rest of the space generally cannot be retrieved. For this reason, in order to obtain optimal results, it is strongly suggested to exploit the detector with quad polarimetric data. However, in the case only dual polarimetric data are available, the algorithm can still be executed as we now show.

The final formal expression of the detector does not suffer significant changes:

$$\gamma_d = \cfrac{1}{\sqrt{1 + RedR\left(\frac{P_{tot}}{P_T} - 1\right)}} = \cfrac{1}{\sqrt{1 + \frac{b^2}{1-2b^2}\left(\frac{\underline{d}^{*T}\underline{d}}{|\underline{d}^{*T}\underline{d}_T|^2} - 1\right)}} \geq T. \qquad (8.23)$$

where the d vector is the dual polarimetric counterpart of $\underline{t}$:

$$\underline{d} = Trace([C_d]\Psi_d) = [t_1, t_2, t_3]^T = \left[\left\langle|k_1|^2\right\rangle, \left\langle|k_2|^2\right\rangle, \left\langle k_1^* k_2\right\rangle\right]^T. \qquad (8.24)$$

$[C_d]$ is a 2×2 coherency matrix calculated starting from the 2 dimensional complex scattering vector for dual polarimetric data.

8.3   Classifier

8.3.1   Formulation

A classifier can be designed starting from the partial target detector, where any class (i.e. partial target) is described by a specific covariance matrix $[C_i]$. The proposed partial target detector is exploited to generate several masks for the specific classes. If only few areas are of interest (e.g. different states of sea ice) a small number of classes are sufficient (the extreme scenario is with one single detection mask). Otherwise, several covariance matrices must be taken into account. The classification output is similar to the supervised Wishart (Cloude and Pottier 1997; Lee et al. 1994a).

The detections of the classes are performed in series generating a stack of masks:

$$\begin{cases} m_i(x, y) = 0 & \text{if } \gamma_i(x, y) < T \\ m_i(x, y) = \gamma_i(x, y) & \text{if } \gamma_i(x, y) \geq T \end{cases} \qquad (8.25)$$

where $i = 1, \ldots, n$ indicates the respective class.

The choice of the SCR for the detectors follows the rationales of generating the class of unknown targets. In the case they are not required, the threshold of the detectors can be set to zero which will lead to a discrimination exclusively on the base of the amplitude of γ_d. In this context, the selection of SCR is trivial and will not affect the final classification mask.

Subsequently, the mask with the maximum value is selected for each pixel. The normalised inner product returning the higher value is the one with the smallest angular distance to the regarded class. If $m_1, \ldots, m_n$ are the n obtained masks, a pixel is allocated to the class Y if:

$$m_Y = \max_{i=1,\dots,n} \{m_i\}. \tag{8.26}$$

In an actual implementation of the classifier, the partial target detector is executed n times one after the other. In any execution, the vector representing the specific class is selected. The classifier is completed by a simple algorithm which pixel by pixel selects the mask presenting the maximum value. The classifier does not require iterations, since it converges after the first attempt.

8.3.2 Parameter Selection

A straightforward strategy could be to simply use the same parameters exploited for standard detection. However, we believe the selection of $SCR = 15$ reveals a significant advantage. As shown by Eq. 8.26, the classifier decision rule is based on the comparison of different masks and selection of the maximum. In this way, the algorithm assigns the pixel to the class with a characteristic vector closer to the observed one. With a lower SCR (i.e. lower selectivity), we are able to detect observed targets presenting some slight dissimilarity from the class characteristic vector. For instance, the dense forest class should include a relatively large collection of volumes (e.g. clouds of particles with different shapes). Clearly, when the difference is too large, a new class must be introduced.

As a general consideration, in the classifier architecture, the use of a detection threshold is exclusively related to the rejection of unknown targets. In the case this is not required, we could choose $SCR = 0$ (which corresponds to $T = 0$) and the discrimination would be performed only by the maximum selection Eq. 8.26.

8.3.3 Supervised and Unsupervised Versions

Depending on the strategy exploited to extract the class coherency matrix, the classifier can be supervised or unsupervised.

The supervised version requires the user interaction for the selection of known areas. This operation can be easily accomplished on a RGB Pauli composite image.

The unsupervised version trains the detector exploiting polarimetric scattering models. A large assortment of models was developed in the past (Cloude 2009). Considering the proposed algorithm represents a general geometrical operation on polarimetric data, any model can be equally exploited. Therefore, it is left to the user to select the most appropriate model for the particular application of interest. We present examples of both supervised and unsupervised detection and classification in the next section.

8.4 Validation of Partial Target Detector

8.4.1 Datasets Employed

In order to provide a large validation of the detector, several datasets with different settings and scenarios were employed.

The first quad polarimetric dataset was acquired by the E-SAR airborne system of DLR (German Aerospace Agency) during the SARTOM campaign in 2006 (Horn et al. 2006). One aim of the campaign was target detection under foliage, for this reason several manmade targets were deployed on open field and under forest $\hat{i}_T$ canopy cover. The frequency band is L and the image has a spatial resolution of 1.1 m in azimuth and about 2 m in range.

Subsequently, a quad-polarimetric L-band ALOS-PALSAR dataset is exploited for the detection of distributed targets. In particular, we consider detection of historical firescars based on their depolarisation behaviour. The images were acquired in Canada close to the town of Manning, Alberta and present a mix of agricultural and forested areas. The pixel size of ALOS quad polarimetric data is around 24×4.5 m (ground range $\times$ azimuth). Moreover, another quad-polarimetric L-band ALOS-PALSAR scene is exploited for a further investigation of land-use classification. The latter was acquired in China in May 2008, close to the city of Taian and the mountain of Culai and represents a mixed urban, agricultural and mountain forest site

The last dataset used is a TerraSAR-X Stripmap dual polarimetric HH/VV acquisition. The represented scene is again Taian in China and the data were acquired in March 2009. The resolution of the sensor is 1.2×6.6 m (range $\times$ azimuth), however the pixel dimension is about 0.9×2.4 m.

With the intention of testing different modalities of the proposed algorithm, the validation is subdivided in separate sections.

8.4.2 Comparison Between Single and Partial Target Detector

Firstly, the ability to detect single targets is examined. The new algorithm is compared with the single target detector [already validated in (Marino et al. 2009, 2010; Marino and Woodhouse 2009)]. Single targets represent a subspace of the partial targets, described by rank one covariance matrices (Cloude 1992). Therefore they are also detectable by the new partial target detector.

In point target detection a high resolution dataset is favourable, therefore the DLR L-band dataset is employed (Horn et al. 2006). Figures 8.1 and 8.2 presents the comparison between the single and partial target detectors. The RGB Pauli image is presented as comparison (Fig. 8.1). In Fig. 8.2, the two algorithms perform similarly, but the resulting masks are not exactly equal. More information is added in the new detector (i.e. the second order statistics of $\underline{k}$) hence slightly better

Fig. 8.1 RGB Pauli image of
the area utilised for
comparison of single (*STD*)
and partial (*PTD*) target
detector

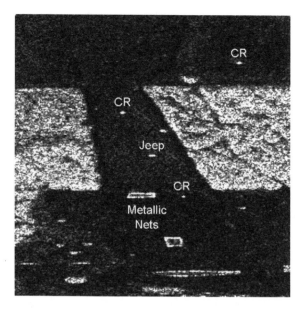

outcomes are expected (i.e. lower false alarm and missed detection rate). The mask
for even-bounces (even number of reflections) identifies mainly the jeep in the
middle of the scene, since it generates a horizontal dihedral with the ground
surface. Moreover it is possible to recognise some trunk-ground double-bounces,
especially on the edge of the forest and on a clearing, where the wave attenuation
due to the canopy is less significant. The masks of odd-bounces (odd number of
reflections) reveal the trihedral corner reflectors and some weaker points on the
bare ground. The metallic nets are rejected since they resemble horizontal dipoles
[as illustrated in (Marino et al. 2009, 2010; Marino and Woodhouse; 2009)].

The capability to reject bright targets is an indicator that the discrimination is
based on the polarimetric information and not the intensity of the return.

8.4.3 Satellite Data: Historical Fire Scar (hfs) Detection

This section is concerned with the exploitation of satellite radar data. The latter are
particularly important for the scientific community and end users since they provide periodical coverage of large areas.

In this section, a quad polarimetric ALOS-PALSAR dataset will be used.
Figure 8.3a and b illustrate respectively the first and third components of the Pauli
scattering vector (i.e. $HH + VV$ and $2*HV$) of a scene acquired in Canada and
presenting a combination of agricultural fields (up left corner) and forests. Considering that the rectangular shape of the pixel introduces severe visual distortions in
the image, the data were multi-looked using an asymmetric window size of 1×5.

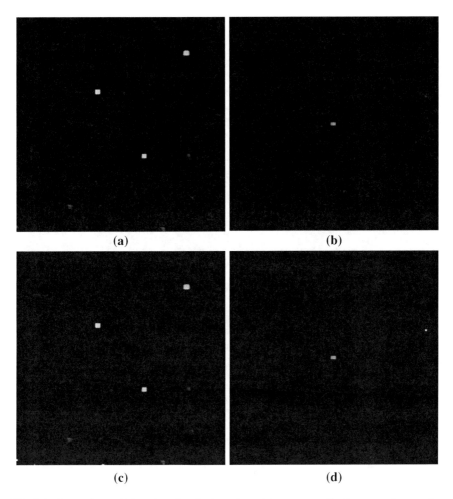

Fig. 8.2 Comparison of detection for single and partial targets. **a** Single target detector for bounces (*STD*); **b** Partial target detector for bounces (*PTD*); **c** Single target detector for dipoles (*STD*); **d** Partial target detector for dipoles (*PTD*)

The multi-look was accomplished on the covariance matrix [*C*] with the intention of preserving the polarimetric information (Lee et al. 1997). The detector uses a subsequent window average of 9 × 9 in order to minimise speckle and accurately characterise depolarised targets in the scene (Lee et al. 1994b).

The test area includes a forest region subject to a fire in 2002 (close to the bottom right corner). The historical fire scar (*hfs*) presents structural differences with the old one due to the younger age of the trees and the absence of understory. Figure 8.4 depicts the detection mask when the algorithm is trained with pixels marked as *hfs* by the ground surveillance. The detector reveals the capability to separate the *hfs* from the rest of the scene, with very low false alarms rate.

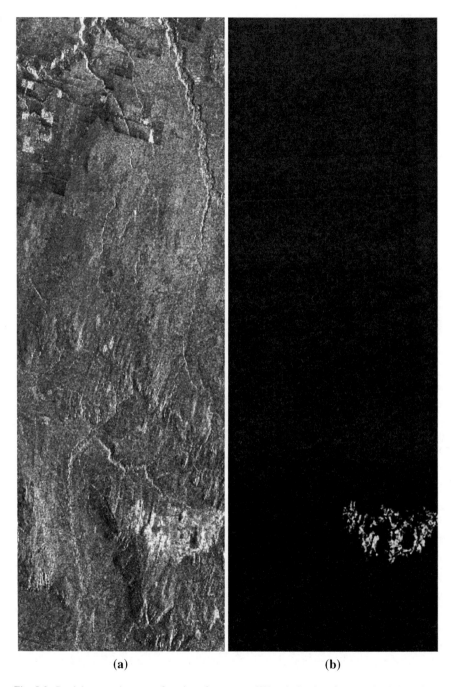

(a)   (b)

Fig. 8.3 Partial target detector of ancient fire scar. **a** HH polarization; **b** supervised detection

The subsequent step considers the examination of a forest model able to link the presence of an *hfs* with some key parameter. The exploited model is the *RVoG* (Random Volume over Ground) (Cloude and Papathanassiou 1998; Treuhaft and Cloude 1999), where the return from the forest is described by random volume scattering plus a coherent component. The latter is commonly generated by the ground beneath the canopy and is described by a rank one coherency matrix (since it is a single target).

The volume contribution is modelled as scattering from dipoles randomly oriented:

$$[T] = [T_S] + [T_V],$$

$$[T_S] = m_S \begin{bmatrix} \cos^2(\alpha) & \cos(\alpha)\sin(\alpha)e^{j\mu} & 0 \\ \cos(\alpha)\sin(\alpha)e^{-j\mu} & \sin^2(\alpha) & 0 \\ 0 & 0 & 0 \end{bmatrix},$$

$$[T_V] = m_V \begin{bmatrix} 2 & 0 & 0 \\ 0 & 1 & 0 \\ 0 & 0 & 1 \end{bmatrix}.$$

(8.27)

Where m_S and m_S are the magnitude of the two backscattering contributions. Their ratio is the ground-to-volume ratio

$$\rho = \frac{m_S}{m_V}.$$

(8.28)

α is the characteristic angle with the same meaning as in the eigenvector decomposition of the coherency matrix $[T]$ (Cloude 2009).

In this experiment, we exploited a model in absence of slopes, since the *DEM* (Digital Elevation Model) of the image is particularly flat, however, in the case of relevant topography a preliminary slope correction should be accomplished (Lee et al. 2002). In order to find the initial values for the model parameters which fit the *hfs*, the model was inverted on the data. The resulting parameters were found to be:

$$\alpha = 19°, \quad \rho = 7.7\,\mathrm{dB} \text{ and } \rho = 3\,\mathrm{dB}.$$

(8.29)

(Similar results, especially regarding α, were found for other *hfs* in Canada). The extracted values were used to reconstruct a $[T]$ matrix to train the detector (Fig. 8.4).

The model seems to approximate adequately the typology of target, since a bad fit would not allow a correct reconstruction of $[T]$. The latter is an example of exploiting a model to train the unsupervised detector, however different models can be employed, such as the oriented volume over ground (*OVoG*) or other multi-layer decompositions (Cloude 2009).

Fig. 8.4 Partial target
detector of ancient fire scar:
unsupervised detection

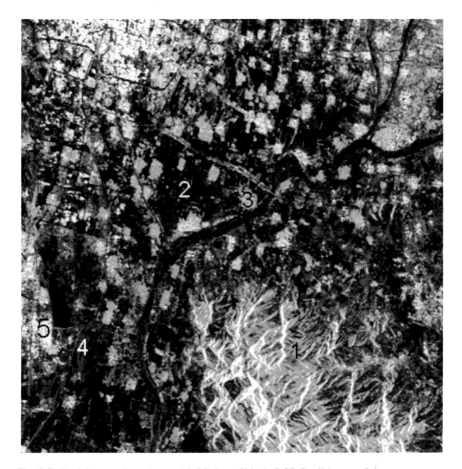

Fig. 8.5 Partial target detection on ALOS data (China): RGB Pauli image of the area

8.4.4 Satellite Data: Classification

In this section the algorithm is evolved into a classifier and tested over a second L-band ALOS-PALSAR dataset in China. The city of Taian (upper left corner) and the mountain of Culai (lower right corner) are clearly visible in the RGB Pauli composite image (Fig. 8.5, where 1200×1200 pixels are visualised here).

Figure 8.6 illustrates the Google Earth image of the areas. The classification mask using the proposed classifier is presented in Fig. 8.7 while its compared with the Wishart supervised (Lee et al. 1999; 1994a) is illustrated in Fig. 8.8. The latter is a classifier exploiting an assumed a priori probability distribution of the coherency matrix [T] (Cloude 2009; Lee et al. 1994a; Lee and Pottier 2009). In this comparison, a basic version of the Wishart supervised classifier was utilised. This is freely available in the software package POLSARpro.

Fig. 8.6 Partial target detection on ALOS data (China): Google Earth photograph of the scene

We are conscious that more elaborated versions employing supplemental pre-processing can result in more accurate classification masks. However, in order to make the comparison as fair as possible, the two classifiers had exactly the same pre-processing and they both are executed in the most basic version. The absence of corrections or further processing should allow us to evaluate anticipated theoretical advantages.

In Fig. 8.5, labels identify the training areas. Area1 represents agricultural fields (blue), Area2 is surfaces (light blue), Area3 is urban area (green), Area4 is a village characterised by small structures and sparse trees (yellow) and Area5 is a dense forest (red). The proposed classification has a total of 6 classes, since the black colour is reserved to areas not falling in any class (i.e. unknown targets). Performing a preliminary detection (setting $SCR = 15$) of the different typologies, the areas are not forced to adhere to any class avoiding misclassification.

The proposed algorithm seems able to separate the different areas in the scene showing significant agreement with the RGB Pauli image the Google Earth photograph. Please note, as in the previous case, the coherency matrix is multi-looked 1×5, however the pixel is not completely square and a distortion of the radar image is still visible. Moreover, the azimuth is not perfectly aligned with the north-south direction. The urban area presents an interesting scenario. The classification mask presents a conspicuous heterogeneity (due to the natural heterogeneity of the

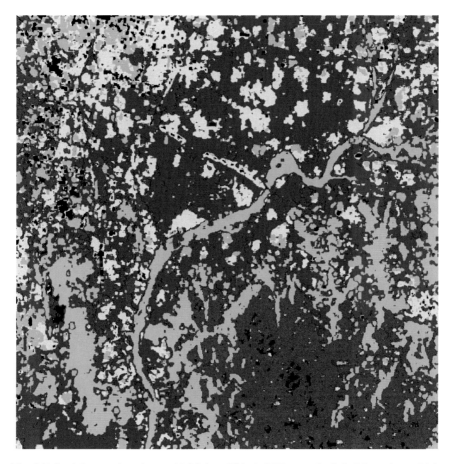

Fig. 8.7 Partial target detection on ALOS data (China): *PTD* supervised. *Red* dense forest, *Light blue* surfaces, *Blue* agricultural, *Yellow* villages, *Green* urban area

city). Specifically, there are several point targets which do not fall in the class and are separated in black. Additionally, the suburban areas resemble more the villages (yellow), rather than the dense city area.

The supervised Wishart classifier (statistical based) (Lee et al. 1999; 1994a) seems to have an overall agreement with the proposed algorithm for two classes: bare surface and agricultural fields. On the other hand, the other areas present rather scarce agreement. Specifically, in Wishart the urban area is much more extended and confused with the villages. For instance, the upper right corner is classified as a town/village while it is an agricultural area.

Moreover, the forest on the mountainous area is completely misclassified presenting a mix of village and urban areas.

From this experiment, a major advantage of the proposed classifier is noticeable: the independence on the total intensity of the backscattering. Wishart is strongly dependent on the Trace of [T] in the calculation of its interclass distance.

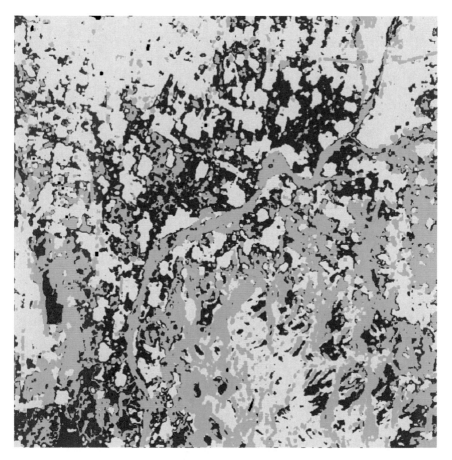

Fig. 8.8 Partial target detection on ALOS data (China): Wishart supervised classifier (same classes as before)

On the other hand, the independence on the overall amplitude focuses our detector exclusively on the polarimetric characteristics (relative weight of the matrix elements). Please note, if in Eq. 8.14 we multiply $[C]$ by a scalar factor the resulting detector does not change. For Wishart, two objects can have a small distance if their power backscattered is similar even though they present some polarimetric difference.

However, in the case the overall amplitude keeps essential physical meanings for a specific target, its information can be taken into account performing a subsequent amplitude analysis over the obtained mask. Nevertheless, the possibility to separate the polarimetric and amplitude information is considered the most significant advantage of the proposed classifier. Obviously, if the effect of amplitude modulation can be corrected with ancillary information (e.g. a DEM) the accuracy of the Wishart supervised classification mask is expected to improve, but such corrections are not always stable and robust and here we have demonstrated an approach that is not so sensitive to errors in topography compensation.

8.4.5 Satellite Data: Dual Polarimetric Detection

In this final experiment, the detector is tested with dual polarimetric data. The basic difference with quad polarimetric data is the lack of uniqueness in the description of an observed target (Cloude 1992). For this reason, an appropriate use of the algorithm should restrict the detection to target typologies which can be represented with sufficient accuracy by only two polarisations. An example is the scattering from a random volume.

In this experiment, TerraSAR-X Stripmap dual polarimetric *HH/VV* data are exploited. As for the ALOS dataset, the scene was acquired in China over the city of Taian. However, now the scene is slightly more north showing the Choushui Xuneng Reservoir (i.e. mountainous area covered by dense forests).

An initial multi-look of 2×3 (azimuth $\times$ range) was performed on the dual polarimetric covariance matrix. Subsequently, the detection was achieved employing a 9×9 boxcar filter. The detection is aimed at volume scattering composed of randomly oriented dipoles. In the case of *HH/VV* dual polarimetry, we do not have direct access to a cross-polarised *HV* channel to detect volume scattering. Instead, the latter can be identified through its signature coherency matrix in the *HH/VV* subspace, expressed as shown in Eq. 8.30:

$$[T_V^d] = m_V \begin{bmatrix} 2 & 0 \\ 0 & 1 \end{bmatrix}. \tag{8.30}$$

Figure 8.9 presents the *HH* reflectivity image compared with the detection mask in Fig. 8.10. The algorithm seems able to identify the mountainous areas covered by dense forest, based on their level of volume scattering. The water reserve, in the middle left of the image, is detected since its backscattering is particularly low and close to the noise floor. Consequently, it resembles random volume with slightly stronger surface component. In order to remove these points, a simple threshold on the amplitude could reject areas with backscattering close to the noise floor. Regarding the detected points externally to the mountainous area, they mainly correspond to trees in the city, besides roads or around fields.

Here, they are more apparent than in the ALOS data due to the enhanced resolution and the use of X band which is more sensitive to canopy. However, we cannot neglect that part of these points are merely false alarm due to the absence of the complete polarimetric information.

Fig. 8.9 Dual polarimetric detector (*HH/VV*) on TerraSAR-X data (China): *HH* reflectivity image

Fig. 8.10 Dual polarimetric detector (*HH/VV*) on TerraSAR-X data (China): detection mask of volume composed of randomly oriented dipoles

8.5 Final Remarks

In this chapter, a geometric interpretation has been provided for the single target detector developed in (Marino et al. 2009, 2010; Marino and Woodhouse 2009) and based on a perturbation filter. The detector is constituted by a weighted (by the observables) and normalised inner product between the target of interest and a perturbed version. In order to extend the detection to partial targets, a new vector formalism was proposed. The new formalism can describe uniquely the partial target space. Finally, the new detector was exploited as first stage of a subsequent classifier.

Validation against airborne (DLR E-SAR, L-band) and satellite data (ALOS-PALSAR and TerraSAR-X) is provided showing the capability of the detector to discriminate among different single and partial targets. The detector is an algebraic operation on a Euclidean space independent of its dimensions. Therefore, a dual polarimetric version can be developed, although we expect lower performances due to the loss of physical information. Both the supervised and unsupervised detection strategies were exploited.

The classification mask is compared with a basic Wishart supervised algorithm (freely available in the software package POLSARpro), revealing what we believe to be a major enhancement: the independence on the overall intensity of the return (i.e. the proposed algorithm works solely with the polarimetric information). Therefore, misclassifications due to modulations of the amplitude, as for example a consequence of layover, are solved, making the new algorithm particularly suited for detection and classification in mountainous regions. Clearly, if ancillary information (as a *DEM*) is available and further pre-processing is performed the classification result of the Wishart supervised can be significantly improved.

Acknowledgments The ALOS PALSAR data for the fire scar detection was provided courtesy of Dr. Hao Chen and Dr. David Goodenough, Canadian Forestry Service (CFS), Victoria, BC. The ALOS-PALSAR data for the Chinese test site was provided courtesy of the DRAGON 2 program. Finally we would like to acknowledge support from TerraSAR-X project number LAN0638 for provision of the dual polarised data used.

References

Cloude RS (1992) Uniqueness of target decomposition theorems in radar polarimetry. Direct and inverse methods in radar polarimetry. Kluwer Academic Publishers, Dordrecht, pp 267–296
Cloude SR (2009) Polarisation: applications in remote sensing. Oxford University Press, Oxford, 978-0-19-956973-1
Cloude SR, Papathanassiou KP (1998) Polarimetric SAR interferometry. IEEE Trans Geosci Remote Sens 36:1551–1565
Cloude SR, Pottier E (1997) An entropy based classification scheme for land applications of polarimetric SAR. IEEE Trans Geosci Remote Sens 35:68–78
Horn R, Nannini M, Keller M (2006) SARTOM airborne campaign 2006: data acquisition report, DLR-HR-SARTOM-TR-001

Huynen JR (1970) Phenomenological theory of radar targets. Delft Technical University, The Netherlands

Lee JS, Pottier E (2009) Polarimetric radar imaging: from basics to applications. CRC Press, Taylor and Francis Group, Boca Raton

Lee JS, Grunes MR, Kwok R (1994a) Classification of multi-look polarimetric SAR imagery based on the complex Wishart distribution. Int J Remote Sens 15:2299–2311

Lee JS, Jurkevich I, Dewaele P, Wambacq P, Oosterlinck A (1994b) Speckle filtering of synthetic aperture radar images: a review. Remote Sens Rev 8:313–340

Lee JS, Grunes MR, Boerner WM (1997) Polarimetric property preserving in SAR speckle filtering. Proc SPIE 3120:236–242

Lee JS, Grunes MR, Ainsworth TL, DU LJ, Schuler DL, Cloude SR (1999) Unsupervised classification using polarimetric decomposition and the complex Wishart classifier. IEEE Trans Geosci Remote Sens 37:2249–2258

Lee JS, Schuler DL, Ainsworth TL, Krogager E, Kasilingam D, Boerner W-M (2002) On the estimation of radar polarization orientation shifts induced by terrain slopes. IEEE Trans Geosci Remote Sens 40:30–41

Marino A , Woodhouse IH (2009) Selectable target detector using the polarization fork. In: IEEE international geoscience and remote sensing symposium IGARSS 2009

Marino A, Cloude SR, Woodhouse IH (2009) Polarimetric target detector by the use of the polarisation fork. In: Proceedings of 4th ESA international workshop, POLInSAR 2009

Marino A, Cloude SR, Woodhouse IH (2010) A polarimetric target detector using the Huynen fork. IEEE Trans Geosci Remote Sens 48:2357–2366

Rose HE (2002) Linear algebra: a pure mathematical approach. Birkhauser, Berlin

Strang G (1988) Linear algebra and its applications, 3rd edn. Thomson Learning, London

Treuhaft RN, Cloude RS (1999) The structure of oriented vegetation from polarimetric interferometry. IEEE Trans Geosci Remote Sens 37:2620–2624

Chapter 9
Conclusions

In the past few decades, radar remote sensing has established itself as an indispensable tool for surveillance, particularly in areas where constant in situ inspections are impracticable (e.g. oceans, deserts, forests, etc.) (Campbel 2007; Chuvieco and Huete 2009). The winning advantage of microwave compared with optical remote sensing is its availability at night time and with any weather conditions, and for longer wavelengths, the ability to penetrate foliage (Richards 2009; Woodhouse 2006). Furthermore, by the use of polarimetry the detected target can be recognised, since different scatterers have different polarimetric responses (Cloude 2009; Lee and Pottier 2009; Mott 2007; Ulaby and Elachi 1990; Zebker and Van Zyl 1991). The latter is the focal point of this thesis.

Firstly, the fundamental concepts of radar polarimetry were illustrated. Polarimetry is a vast subject ranging from physics to algebra (Born and Wolf 1965; Cloude 1995, 2009). A thorough treatment of polarimetry was outside the scope of this thesis and we decided to include exclusively the concepts directly related to the algorithm development. Specifically, the possibility to describe a deterministic target with a scattering matrix or equivalently a scattering vector was described. The latter is a three dimensional complex vector, therefore any target of interest can be pictured with a vector in a three dimensional space. This algebraic abstraction is very powerful and represents the basis of the detector developed here (Cloude 1995).

After a preliminary introduction, the novel polarimetric detector was developed (Marino et al. 2009, 2010; Marino and Woodhouse 2009). The algorithm is based on a polarimetric coherence between the target of interest and its perturbed version (i.e. a slightly rotated vector). The coherence was calculated as a weighted and normalised inner product, where the weights are extracted from the data. A threshold on the coherence finished the detection algorithm. The mathematical formalism obtained is straightforward and the numerical calculations required a remarkably short processing time. Starting from the mathematical expression, the detector parameters were optimised in order to have unbiased detections, when a priori information about the clutter are absent.

A. Marino, *A New Target Detector Based on Geometrical Perturbation Filters for Polarimetric Synthetic Aperture Radar (POL-SAR)*, Springer Theses, DOI: 10.1007/978-3-642-27163-2_9, © Springer-Verlag Berlin Heidelberg 2012

Subsequently, the evaluation of the theoretical detector performance was treated in detail exploiting the statistics of the coherence (i.e. detector) interpreted as a random variable (Monahan 2001; Papoulis 1965). In order to describe completely the statistical behaviour, the analytical probability density function (*pdf*) was calculated. The theoretical results, and in particular the Receiver Operating Characteristic (ROC) curves revealed that the detector performance is particularly high with an almost deterministic behaviour (Kay 1998). Subsequently, the theoretical performance was compared with other detectors. Under similar detection conditions, the ROC of the proposed detector appeared to highly outperform other algorithms (e.g. Polarimetric Whitening Filter, Optimum Polarimetric Detector, etc.) (Chaney et al. 1990; Novak et al. 1993). The excellent performance is a consequence of the strong reduction of the random part of the detector performed by the coherence formalism. Finally, the detection performance remains high as long as the polarimetric description of the target of interest is accurate (please note, the target description can be attained from a theoretical model or extracted from a dataset). Once the statistical description is available, the threshold on the coherence can be chosen optimally (Kay 1998). Considering the variance is particularly small the optimisation process is relatively straightforward. A simple graphical procedure was proposed in this thesis.

Finally, the optimised detector was validated on real data. In this thesis, airborne and satellite data were treated separately since the former represent an easier detection scenario due to enhanced spatial resolution and Signal to Noise Ratio (SNR). Specifically, the airborne data (E-SAR from DLR) were collected in the framework of the SARTOM project (Horn et al. 2006). The latter was designed for target detection under foliage (FOLPEN) with polarimetric and tomographic SAR (Walker et al. 2010). Due to the presence of deterministic targets (e.g. corner reflectors, jeeps, containers, etc.) deployed in open field and under foliage, this dataset represents an ideal scenario for target detection. With the purpose of providing a wider validation (not exclusively restricted to deployed targets), several pictures of the test site were collected, especially in areas where the algorithm was showing positive detections.

In order to understand if the theoretical improvements indicated by the ROC curves are reflected in the actual target detection, the algorithm was compared with a widely used detector, the Polarimetric Whitening Filter (PWF). The latter was defined as the best signal processing for speckle reduction (Chaney et al 1990; Novak et al 1993; Novak and Hesse 1993). The proposed new detector was demonstrated to outperform the PWF in the established criteria for the evaluation of the algorithm performance (as detailed in the following).

The eighth chapter concerned the validation with satellite data. Specifically, ALOS-PALSAR (L-band), RADARSAT-2 (C-band) and TerraSAR-X (X-band) were exploited. In this case, we could not utilise accurate ground truth of the test areas, therefore to help the validation we compared the masks with aerial and ground photographs provided by Google Earth and Panoramio. The algorithm revealed its capability to perform detection with satellite data, opening the possibility to constantly monitor wide regions without the need for airborne campaigns.

In the last chapter, the single target detector was converted to a partial (de-polarised) target detector. In this context, a formalism utilising the coherence matrix is necessary, since the scattering matrix alone is not sufficient to completely describe a partial target. The new algorithm was built with the same perturbation filter procedure. With a partial target detector it is possible to detect any polari-metric target with large advantages in classification for land use monitoring more than surveillance (Marino et al. 2012). The performance of the new classifier is tested on real data (E-SAR, ALOS PALSAR and TerraSAR X) and compared with a supervised Wishart classifier.

The wide validation over real data and comparison with other detectors allow a revision of the criteria set for the evaluation of the algorithm performance. Spe-cifically, the criteria were based on two probabilities:

(1) Low probability of missing a target on the scene (i.e. missed detection). As shown in the fifth chapter, in the case of ordinary detection (i.e. $SCR > 2$) the theoretical probability of missed detection (i.e. P_M) is extremely small and negligible when the threshold is selected with the proposed graphical proce-dure. For this reason, the criterion can be considered largely fulfilled from the theoretical point of view. However, we want to provide a clarification in order to identify the limits of the algorithm. The criterion is completely fulfilled as long as the polarimetric signature of the target of interest is accurate. Intui-tively, a wrong signature can lead to misidentification of the target. Two different procedures can be employed to acquire the target signature. The first one is model based and is to be preferred in the case the target is relatively easy to model. The second one considers the extraction (or training) from another dataset (but with same characteristics). In order to mitigate the effects of errors in the target signature, the algorithm can detect real targets slightly different from the one of interest. Mathematically, this is expressed by a dispersion equation where the limits can be set by the user.

Two main targets typologies were identified as particularly interesting:

(1.1) **Targets under foliage cover**. Detection under foliage (FOLPEN) is a topic of remarkable interest for military and civilian surveillance since patrolling forested areas with ground inspections is highly difficult (Fleischman et al. 1996). Furthermore, the inspection is particularly complicated with optical systems because the tree canopy generally represents a barrier for optical sensors, while microwave can partially penetrate providing information about targets on the ground. The enhanced performance of the new method (compared with PWF) is due to the different use of the polarimetric infor-mation. PWF assumes that artificial targets do not present speckle variation, hence speckled points are rejected (Novak et al. 1993). On the other hand, the proposed detector is concerned with the exact polarimetric signature of the target (it does not consider the variance of the observables). In foliage penetration, the canopy can introduce speckle in the return from the area where the target is located.

(1.2) **Small targets**. The detector is based on a polarimetric coherence (normalised operator) rather than the amplitude of the backscattering (or *Trace* of
   the coherency matrix). This represents a significant advantage in detection of
   targets with a small radar cross section (low backscattering), outperforming
   algorithms based on thresholds on the amplitude (Chaney et al. 1990; Li and
   Zelnio 1996). Clearly, the target still has to be stronger than the background
   (or the noise level) to be detected (otherwise we would detect random noise,
   unless we have a priori information about the clutter).

(2) Low probability of positive detection in absence of an actual target (i.e. false
   alarm):

(2.1) **Statistical stability**. The theoretical results obtained in Chap. 5 clearly
   revealed that in the case of ordinary detection (i.e. $SCR > 2$) the probability
   of false alarm (i.e. P_F) is negligible. Furthermore, the proposed algorithm
   largely outperformed all the other detectors considered in the comparison
   (Chaney et al. 1990). Therefore, we consider this point strongly fulfilled.

(2.2) **Robustness against bright natural targets**. Another advantage of neglecting the amplitude is when there exist several natural targets with strong
   backscatter. In some cases, a big radar cross section can be completely unre-
   lated to the target itself but merely due to the acquisition geometry. Areas in
   layover (the side of a mountain or the edge of a forest) can have an extraordinary strong return that can trigger an amplitude based detector. On the other
   hand, the polarimetric information remains quite stable in the presence of
   layover. In any case, a simple correction can be performed as pre-processing on
   the data which takes into account possible changes of polarimetric characteristics due to slopes (Lee et al. 2002). Please note, the correction does not
   require any a priori information (e.g. DEM) and does not change the total
   amplitude of the return. Finally, we can consider the criterion fulfilled.

A final advantage of the proposed detector not included in the initial criteria is
the simplicity of the final mathematical expression. This allows particularly fast
processing. The algorithm was able to execute the detection over all the dataset
(about 2 million pixels) in few seconds, making it feasible for near real time
processing of the data (given suitable optimisation).

References

Born M, Wolf E (1965) Principles of Optics, 3rd edn. Pergamon Press, New York
Campbel JB (2007) Introduction to remote sensing. The Guilford Press, New York
Chaney RD, Bud MC, Novak LM (1990) On the performance of polarimetric target detection
   algorithms. IEEE Aerosp Electron Syst Mag 5:10–15
Chuvieco E, Huete A (2009) Fundamentals of satellite remote sensing. Taylor & Francis Ltd,
   New York
Cloude SR (1995) Lie groups in EM wave propagation and scattering. Chapter 2 in
   Electromagnetic symmetry.Baum C, Kritikos HN (eds) Taylor and Francis, Washington,
   pp 91–142. ISBN 1-56032-321-3

Cloude SR (2009) Polarisation: applications in remote sensing. Oxford University Press, 978-0-19-956973-1

Fleischman JG, Ayasli S, Adams EM (1996) Foliage attenuation and backscatter analysis of SAR imagery. IEEE Trans.Aerosp Electron Syst Mag 32:135–144

Horn R, Nannini M, Keller M (2006) SARTOM airborne campaign 2006: data acquisition report. DLR-HR-SARTOM-TR-001

Kay SM (1998) Fundamentals of statistical signal processing, vol 2: detection theory, Prentice Hall, Lynnfield

Lee JS, Pottier E (2009) Polarimetric radar imaging: from basics to applications. CRC Press, Boca Raton

Lee J, Schuler DL, Ainsworth TL, Krogager E, Kasilingam D, Boerner WM (2002) On the estimation of radar polarization orientation shifts induced by terrain slopes. IEEE Trans Geosci Remote Sens 40(1):30–41

Li J, Zelnio EG (1996) Target detection with synthetic aperture radar. IEEE Trans. Aerosp Electron Syst Mag 32:613–627

Marino A, Woodhouse IH (2009) Selectable target detector using the polarization fork. In: IEEE international geoscience and remote sensing symposium IGARSS 2009

Marino A, Cloude S, Woodhouse IH (2009) Polarimetric target detector by the use of the polarisation fork. In: Proceedings of 4th ESA international workshop, POLInSAR 2009

Marino A, Cloude SR, Woodhouse IH (2010) A polarimetric target detector using the huynen fork. IEEE IEEE Trans. Geosci. Remote Sens 48:2357–2366

Marino A, Cloude SR, Woodhouse IH (2012) Detecting depolarizing targets using a new geometrical perturbation filter. IEEE Trans Geosci Remote Sens (Next available issue)

Monahan JF (2001) Numerical methods of statistics. Cambridge University Press, Cambridge

Mott H (2007) Remote sensing with polarimetric radar. Wiley, Hoboken

Novak LM, Hesse SR (1993) Optimal polarizations for radar detection and recognition of targets in clutter. In: Proceedings, IEEE national radar conference, Lynnfield, pp 79–83

Novak LM, Burl MC, Irving MW (1993) Optimal polarimetric processing for enhanced target detection. IEEE Trans.Aerosp Electron Syst Mag 20:234–244

Papoulis A (1965) Probability, random variables and stochastic processes. McGraw-Hill, New York

Richards JA (2009) Remote sensing with imaging radar-signals and communication technology. Springer-Verlag Berlin and Heidelberg GmbH & Co K, Berlin

Ulaby FT, Elachi C (1990) Radar polarimetry for geo-science applications. Artech House, Norwood

Walker N, Horn R, Marino A, Nannini M, Woodhouse IH (2010) The SARTOM project: tomography and polarimetry for enhanced target detection for foliage penetrating airborne P-band and L-band SAR. In: EMRS-DTC 2008, 6th Annual Technical Conference, Edinburgh, 13–14 July

Woodhouse IH (2006) Introduction to microwave remote sensing. CRC press, Taylor & Frencies Group, New York

Zebker H, Van Zyl JJ (1991) Imaging radar polarimetry. A review. Proc IEEE 79(11):1583–1606

Appendix 1
Geometrical Perturbation
with the Huynen Parameters

In the development of the polarimetric detector, the geometrical perturbation imposed to the target vector plays a fundamental role. In Chap. 4, the procedure was generally described introducing three equivalent approaches. In this appendix, a more rigorous formulation will be provided exploiting the Huynen parameterisation (this is preferred for its narrow link with phenomenological properties of the target) (Huynen 1970). Please note, the same results can be obtained using the α model (or any parameterisation based on continuous functions) (Cloude 2009, Cloude and Pottier 1997).

A generic target can be described by the Huynen parameters as

$$[S_T] = [R(\psi_m)][T(\chi_m)][S_d(\gamma, \upsilon)][T(\chi_m)][R(-\psi_m)],$$

$$[S_d] = \begin{pmatrix} e^{i\upsilon} & 0 \\ 0 & \tan(\gamma)e^{-i\upsilon} \end{pmatrix},$$

$$[T(\chi_m)] = \begin{pmatrix} \cos\chi_m & -i\sin\chi_m \\ -i\sin\chi_m & \cos\chi_m \end{pmatrix}, \tag{1.1}$$

$$[R(\phi_m)] = \begin{pmatrix} \cos\psi_m & -\sin\psi_m \\ \sin\psi_m & \cos\psi_m \end{pmatrix}.$$

In Eq. 1.1, the value of m is set to 1 since we are generating a scattering mechanism, and the absolute phase is neglected (or $\zeta = 0$). Starting from a normalised scattering matrix, the scattering mechanism can be obtained in the classical way (Lee and Pottier 2009):

$$\underline{\omega}_T(\psi_m, \chi_m, \upsilon, \gamma) = \frac{1}{2} Trace([S_T]\Psi). \tag{1.2}$$

The geometrical perturbation is performed by changing slightly the value of the Huynen parameters representing the target of interest. The perturbed target is

$$\underline{\omega}_P(\psi_m \pm \Delta\psi_m, \tau_m \pm \Delta\tau_m, \upsilon \pm \Delta\upsilon, \gamma \pm \Delta\gamma), \tag{1.3}$$

A. Marino, *A New Target Detector Based on Geometrical Perturbation Filters for Polarimetric Synthetic Aperture Radar (POL-SAR)*, Springer Theses, DOI: 10.1007/978-3-642-27163-2, © Springer-Verlag Berlin Heidelberg 2012

where $\Delta\psi_m, \Delta\tau_m, \Delta\upsilon$ and $\Delta\gamma$ are the real positive variations of the parameters. They correspond to a fraction of the respective parameter variation range (i.e. $\psi_m \in [0, \pi]$, $\tau_m \in [-\pi/4, \pi/4]$, $\gamma \in [0, \pi/4]$ and $\upsilon \in [-\pi/2, \pi/2]$).

We want to demonstrate:

if the variations $\Delta\psi_m$, $\Delta\tau_m$, $\Delta\upsilon$ and $\Delta\gamma$ are small (compared with the total range of variation) then $\underline{\omega}_P \approx \underline{\omega}_T$.

(1) First it will be demonstrated:

if the Huynen parameters are changed slightly, the scattering matrix changes slightly.

All the functions exploited in the Huynen parameterisation are continuous in the given intervals. Therefore, it is always possible to examine a small interval where the function is approximately linear (Riley et al. 2006). For small variations we can write:

$$\sin(\psi_m \pm \Delta\psi_m) \approx \sin(\psi_m) \pm c_s\Delta\psi_m, \tag{1.4}$$

where $0 \le c_s(\psi_m) \le 1$ is a real factor equal to the derivative of $\sin(\psi_m)$:

$$c_s(\psi_m) = \frac{d}{d\psi_m}\sin(\psi_m) = \cos(\psi_m). \tag{1.5}$$

Remarkably, $c_s\Delta\psi_m$ is always not bigger than $\Delta\psi_m$ since c_s is always not bigger than 1. Please note, the derivative can be positive or negative, therefore the final sign of the variation depends on the sign of the local derivative.

Regarding the cosine:

$$\cos(\psi_m \pm \Delta\psi_m) \approx \cos(\psi_m) \pm c_c\Delta\psi_m, \tag{1.6}$$

where

$$0 \le c_c(\psi_m) \le 1,$$
$$c_c(\psi_m) = \frac{d}{d\psi_m}\cos(\psi_m) = -\sin(\psi_m). \tag{1.7}$$

Finally, for the tangent function we can write:

$$\tan(\gamma \pm \Delta\gamma) \approx \tan(\gamma) \pm c_t\Delta\gamma, \tag{1.8}$$

where

$$1 \le c_t(\gamma) \le 2,$$
$$c_t(\gamma) = \frac{d}{d\gamma}\tan(\gamma) = \frac{1}{\cos^2(\gamma)}, \tag{1.9}$$

because $0 \le \gamma \le \pi/4$ and the derivative varies between one and two. In the worst scenario, the value of $\Delta\gamma$ can be doubled (depending on the local derivative), however it remains constrained and does not amplify excessively.

Once the variation of the Huynen parameters is performed, the perturbed target can be expressed as

$$[S_T] = [R(\psi_m + \Delta\psi_m)][T(\chi_m + \Delta\chi_m)][S_d(\gamma + \Delta\gamma, \upsilon + \Delta\upsilon)]$$
$$\times [T(\chi_m + \Delta\chi_m)][R(-(\psi_m + \Delta\psi_m))] \quad (1.10)$$

Next step evaluates how the single matrices change when the new parameters are substituted (Riley et al. 2006).

$$[R(\psi_m + \Delta\psi_m)]$$

$$= \begin{bmatrix} \cos(\psi_m + \Delta\psi_m) & -\sin(\psi_m + \Delta\psi_m) \\ \sin(\psi_m + \Delta\psi_m) & \cos(\psi_m + \Delta\psi_m) \end{bmatrix} = \begin{bmatrix} \cos\psi_m \pm c_c\Delta\psi_m & -(\sin\psi_m \pm c_c\Delta\psi_m) \\ \sin\psi_m \pm c_c\Delta\psi_m & \cos\psi_m \pm c_c\Delta\psi_m \end{bmatrix}$$

$$= \begin{bmatrix} \cos\psi_m & -\sin\psi_m \\ \sin\psi_m & \cos\psi_m \end{bmatrix} + \begin{bmatrix} \pm c_c\Delta\psi_m & \mp c_c\Delta\psi_m \\ \pm c_c\Delta\psi_m & \pm c_c\Delta\psi_m \end{bmatrix}$$

$$= [R(\psi_m)] + [\overline{R}(\Delta\psi_m)] \quad (1.11)$$

$$[T(\chi_m + \Delta\chi_m)] = \begin{bmatrix} \cos(\chi_m + \Delta\chi_m) & -j\sin(\chi_m + \Delta\chi_m) \\ -j\sin(\chi_m + \Delta\chi_m) & \cos(\chi_m + \Delta\chi_m) \end{bmatrix}$$

$$= \begin{bmatrix} \cos\chi_m \pm c_c\Delta\chi_m & -j(\sin\chi_m \pm c_c\Delta\chi_m) \\ -j(\sin\chi_m \pm c_c\Delta\chi_m) & \cos\chi_m \pm c_c\Delta\chi_m \end{bmatrix}$$

$$= \begin{bmatrix} \cos\chi_m & -j(\sin\chi_m) \\ -j(\sin\chi_m) & \cos\chi_m \end{bmatrix} + \begin{bmatrix} \pm c_s\Delta\chi_m & \mp j(c_s\Delta\chi_m) \\ \mp j(c_s\Delta\chi_m) & \pm c_s\Delta\chi_m \end{bmatrix}$$

$$= [T(\chi_m)] + [\overline{T}(\Delta\chi_m)] \quad (1.12)$$

$$[S_d(\gamma + \Delta\gamma, \upsilon + \Delta\upsilon)] = \begin{bmatrix} e^{j(\upsilon + \Delta\upsilon)} & 0 \\ 0 & \tan(\gamma + \Delta\gamma)e^{-j(\upsilon + \Delta\upsilon)} \end{bmatrix}$$

$$= \begin{bmatrix} e^{j\upsilon} & 0 \\ 0 & \tan(\gamma + \Delta\gamma)e^{-j\upsilon} \end{bmatrix} \begin{bmatrix} e^{j\Delta\upsilon} & 0 \\ 0 & e^{-j\Delta\upsilon} \end{bmatrix}$$

$$= \left(\begin{bmatrix} e^{j\upsilon} & 0 \\ 0 & \tan\gamma\, e^{-j\upsilon} \end{bmatrix} + \begin{bmatrix} 0 & 0 \\ 0 & c_t\Delta\gamma\, e^{-j\upsilon} \end{bmatrix} \right) \begin{bmatrix} e^{j\Delta\upsilon} & 0 \\ 0 & e^{-j\Delta\upsilon} \end{bmatrix}$$

$$\quad (1.13)$$

If the value of $\Delta\upsilon$ is small enough we can approximate the exponential with the first order of the Taylor series (Riley et al. 2006; Mathews and Howell 2006):

$$e^x \approx 1 + x. \quad (1.14)$$

$$
\begin{aligned}
&\left(\begin{bmatrix} e^{jv} & 0 \\ 0 & \tan\gamma\, e^{-jv} \end{bmatrix} + \begin{bmatrix} 0 & 0 \\ 0 & c_t \Delta\gamma\, e^{-jv} \end{bmatrix}\right) \begin{bmatrix} 1+j\Delta v & 0 \\ 0 & 1-j\Delta v \end{bmatrix} \\
&= \left(\begin{bmatrix} e^{jv} & 0 \\ 0 & \tan\gamma\, e^{-jv} \end{bmatrix} + \begin{bmatrix} 0 & 0 \\ 0 & c_t \Delta\gamma\, e^{-jv} \end{bmatrix}\right) \left(\begin{bmatrix} 1 & 0 \\ 0 & 1 \end{bmatrix} + \begin{bmatrix} j\Delta v & 0 \\ 0 & -j\Delta v \end{bmatrix}\right) \\
&= \left[S_d(\gamma, v) \right] + \left[\overline{S}_d(\gamma, v, \Delta v) \right] + \left[\overline{\overline{S}}_d(\Delta\gamma, v) \right] + \left[\overline{\overline{\overline{S}}}_d(\Delta\gamma, v, \Delta v) \right]
\end{aligned}
$$

$$(1.15)$$

After these passages, the matrices of the perturbed target can be expressed as the superposition of a non-perturbed matrix (i.e. the target of interest) and another matrix linearly dependent on the variation (Pearson 1986; Riley et al. 2006):

$$[B_P] = [B_T] + [B_\Delta], \tag{1.16}$$

where $[B]$ represents any matrix in the parameterisation. $[B_P]$ is the perturbed matrix, $[B_T]$ is the original matrix (non-perturbed) and $[B_\Delta]$ is the variation matrix (linearly dependent on the variation).

The matrices dependent on the variations vanishes (they are equal to the Null matrix), when the variations are null (Hamilton 1989; Rose 2002; Strang 1988). In other words, if $\Delta\phi_m = \Delta\tau_m = \Delta v = \Delta\gamma = 0$, then

$$\left[\overline{T}(\Delta\chi_m) \right] = \left[\overline{R}(\Delta\psi_m) \right] = \left[\overline{S}_d(\gamma, v, \Delta v) \right] = \left[\overline{\overline{S}}_d(\Delta\gamma, v) \right] = \left[\overline{\overline{\overline{S}}}_d(\Delta\gamma, v, \Delta v) \right] = [0],$$

$$(1.17)$$

where the null matrix is defined as:

$$[0] = \begin{bmatrix} 0 & 0 \\ 0 & 0 \end{bmatrix}. \tag{1.18}$$

This proves that the derived expression collapses into the target of interest when the variations are zero.

Coming back to the total expression of the perturbed target:

$$
\begin{aligned}
[S_P] &= \left([R(\psi_m)] + [\overline{R}(\Delta\psi_m)] \right)\left([T(\chi_m)] + [\overline{T}(\Delta\chi_m)] \right) \\
&\times \left([S_d(\gamma, v)] + [\overline{S}_d(\gamma, v, \Delta v)] + [\overline{\overline{S}}_d(\Delta\gamma, v)] + [\overline{\overline{\overline{S}}}_d(\Delta\gamma, v, \Delta v)] \right) \\
&\times \left([T(\chi_m)] + [\overline{T}(\Delta\chi_m)] \right)\left([R(-\psi_m)] + [\overline{R}(\Delta\psi_m)] \right)
\end{aligned}
\tag{1.19}
$$

The product generates several terms, each of them is composed by the multiplication of 5 matrices. The distributive property of the matrix multiplication can be employed to calculate the final expression (Strang 1988):

$$([A] + [B])[C] = [A][C] + [B][C]. \tag{1.20}$$

The final result can be summarised in

$$[S_P] = [S_T] + [A_1] + [A_2] + [A_3] + [A_4] + [A_5], \qquad (1.21)$$

where $[A_1]$, $[A_2]$, $[A_3]$, $[A_4]$ and $[A_5]$ are matrices obtained summing up matrices where there are respectively 1, 2, 3, 4 and 5 variation matrices (hence they vanish when $\Delta = 0$). Please note, the $[A]$ matrices are the sum of several matrices, where only $[A_5]$ is composed exclusively of one addend. The term which has the biggest effect in changing the perturbed target is $[A_1]$, while the others can be interpreted as second order contributions (Mathews and Howell 2006; Riley et al. 2006).

As introduced before when $\Delta\psi_m = \Delta\chi_m = \Delta\upsilon = \Delta\gamma = 0$ then we have

$$[A_1] = [A_2] = [A_3] = [A_4] = [A_5] = [0]. \qquad (1.22)$$

When $\Delta\psi_m$, $\Delta\chi_m$, $\Delta\upsilon$ and $\Delta\gamma$ are small, the $[A_1]$, $[A_2]$, $[A_3]$, $[A_4]$ and $[A_5]$ matrices (especially $[A_1]$) start to be different from the null matrix, thus changing the value of $[S_P]$. Considering the linear dependence of the variation matrices $[B_\Delta]$ on the respective variations, they can be close as we like to the null matrices. Therefore, the perturbed matrix can be close as we like to the target of interest.

(2) The detector formulation is based on the scattering vector formalism and employs a change of basis which makes $\underline{\omega}_T = [1, 0, 0]^T$. In this second section we want to demonstrate:

if the perturbation is applied on the basis which makes $\underline{\omega}_T = [1, 0, 0]^T$, the first component (i.e. target) of the perturbed target is slightly reduced and the other two (i.e. clutters) are increased, leading to a final expression $\underline{\omega}_T = [a, b, c]^T$ with a, b and c complex numbers and $|a| \gg |b|$, $|a| \gg |c|$.

In order to obtain $\underline{\omega}_T = [1, 0, 0]^T$, we can perform two real rotations and one change of phase (please note only one change of phase is required because the target vector has only one component) (Cloude 1995a; Strang 1988). The first rotation deletes one component (or locates the vector on a complex plane orthogonal to one component), and the second overlap the vector on one axis. The change of phase erases the phase of the vector and makes it a real number. Furthermore, the last change of phase can be neglected in this context since it can be assimilated to the final absolute phase (and in any case it does not change the weights of $|a|$, $|b|$ and $|c|$).

The rotations can be accomplished left multiplying the vector for a unitary matrix (Rose 2002). If $\underline{b}_T$ is the given scattering mechanism of the target to detect (expressed in any basis):

$$[U]\underline{b}_T = \underline{\omega}_T, \qquad (1.23)$$

where $[U]$ is a unitary matrix computed as:

$$[U] = \begin{bmatrix} 1 & 0 & 0 \\ 0 & \cos\sigma & -\sin\sigma \\ 0 & \sin\sigma & \cos\sigma \end{bmatrix} \begin{bmatrix} \cos\varphi & 0 & \sin\varphi \\ 0 & 1 & 0 \\ -\sin\varphi & 0 & \cos\varphi \end{bmatrix}$$

$$= \begin{bmatrix} \cos\varphi & 0 & \sin\varphi \\ \sin\sigma\sin\varphi & \cos\phi & -\sin\sigma\cos\varphi \\ -\cos\sigma\sin\varphi & \sin\phi & \cos\sigma\cos\varphi \end{bmatrix} \tag{1.24}$$

Hence,

$$\begin{bmatrix} \cos\varphi & 0 & \sin\varphi \\ \sin\sigma\sin\varphi & \cos\sigma & -\sin\sigma\cos\varphi \\ -\cos\sigma\sin\varphi & \sin\sigma & \cos\sigma\cos\varphi \end{bmatrix} \begin{bmatrix} a' \\ b' \\ c' \end{bmatrix} = \begin{bmatrix} 1 \\ 0 \\ 0 \end{bmatrix} \tag{1.25}$$

In order to calculate the appropriate rotation angles the following system of equations must be solved

$$\begin{cases} a'\cos\varphi + c'\sin\varphi = 1 \\ a'\sin\sigma\cos\varphi + b'\cos\sigma - c'\sin\sigma\cos\varphi = 0 \\ -a'\cos\sigma\sin\varphi + b'\sin\sigma + c'\cos\sigma\cos\varphi = 0 \end{cases} \tag{1.26}$$

The solutions are:

$$\begin{cases} \varphi = \tan^{-1}\left(\frac{1-a'}{c'}\right) \\ \sigma = \tan^{-1}\left(\dfrac{b'}{c'\cdot\cos\left(\tan^{-1}\left(\frac{1-a'}{c}\right)\right) - a'\cdot\sin\left(\tan^{-1}\left(\frac{1-a'}{c}\right)\right)}\right) \end{cases} \tag{1.27}$$

Substituting the values of φ and σ in the expression of $[U]$ the desired change of basis is achieved.

The same change of basis must be applied on the perturbed target $\underline{b}_P = \left[a'', b'', c''\right]^T$:

$$[U]\underline{b}_P = \underline{\omega}_T, \tag{1.28}$$

where

$$\begin{cases} a''\cos\varphi + c''\sin\varphi = a \\ a''\sin\sigma\cos\varphi + b''\cos\sigma - c''\sin\sigma\cos\varphi = b \\ -a''\cos\sigma\sin\varphi + b''\sin\sigma + c''\cos\sigma\cos\varphi = c \end{cases} \tag{1.29}$$

$a'', b'', c'' \neq a', b', c'$ since the perturbed target is different from the target of interest in any basis (hence, in the starting basis). On the other side the values for φ and σ are the same used previously, consequently the triplet a, b, c cannot be 1, 0, 0. However, considering the system is linear, if the change is small enough

$[a'',b'',c'']^T$ will not be very different from $[a',b',c']^T$ in the starting basis (as demonstrated in the first part) and $|a| \approx 1$, $|b| \approx 0$, $|c| \approx 0$.

Finally, we have demonstrated that for small changes of the variation parameters $\Delta\psi_m$, $\Delta\chi_m$, Δv and $\Delta\gamma$, the perturbed target remains similar to the one of interest in any basis and in particular in the basis which makes $\underline{\omega}_T = [1,0,0]^T$.

The entire proof can be summarised in two equations:

$$\underline{\omega}_T = \frac{1}{2}[U(\varphi,\sigma,\delta)]Trace([S_T(\psi_m,\chi_m,\gamma,v)]\Psi) = [1,0,0]^T, \qquad (1.30)$$

where the brackets show the dependence on the Huynen parameters. Again, the perturbed target is obtained changing slightly the Huynen parameters:

$$\underline{\omega}_P = \frac{1}{2}[U(\varphi,\sigma,\delta)]Trace([S(\psi_m + \Delta\psi_m, \chi_m + \Delta\chi_m, \gamma + \Delta\gamma, v + \Delta v)]\Psi)$$
$$= [a(\Delta\psi_m,\Delta\chi_m,\Delta\gamma,\Delta v), b(\Delta\psi_m,\Delta\chi_m,\Delta\gamma,\Delta v), c(\Delta\psi_m,\Delta\chi_m,\Delta\gamma,\Delta v)]^T$$
$$(1.31)$$

If the variation is zero the perturbed target is exactly the target to detect:

$$\Delta\psi_m = \Delta\chi_m = \Delta v = \Delta\gamma = 0 \Leftrightarrow \underline{\omega}_T = \underline{\omega}_P. \qquad (1.32)$$

On the other hand, if the variations are small the two scattering mechanisms start to be different, introducing the required distance:

$$\Delta\psi_m \approx \Delta\chi_m \approx \Delta v \approx \Delta\gamma \approx 0 \Leftrightarrow \underline{\omega}_T \approx \underline{\omega}_P. \qquad (1.33)$$

Geometrically the change of the Huynen parameters results in a small rotation of the perturbed scattering mechanism, introducing an angular distance between them. Considering the target of interest is present only in the first component, this rotation introduces two clutter terms in the perturbed scattering mechanism. As demonstrated in Chap. 4, the possibility to adjust properly the weight between target and these two clutter terms is an essential aspect in the detection algorithm.

Appendix 2
Neglecting the Cross Terms

The polarimetric detector developed in this thesis is constructed with a weighted and normalised inner product between the vector representing the target to detect $\underline{\omega}_T$ and a perturbed replica $\underline{\omega}_P$. The weights are derived from the observables (scattering vector). Specifically, the matrix representing the weights is defined as:

$$[A] = \begin{bmatrix} k_1 & 0 & 0 \\ 0 & k_2 & 0 \\ 0 & 0 & k_3 \end{bmatrix}, \tag{2.1}$$

where the scattering vector is $\underline{k} = [k_1, k_2, k_3]^T$ (Cloude 2009; Lee and Pottier 2009). The inner product can be calculated as

$$\underline{\omega}_T^{*T}[P]\underline{\omega}_T, \tag{2.2}$$

where the $[P]$ matrix can be calculated with the Hermitian product of $[A]$:

$$[P] = [A]^{*T}[A] = \begin{bmatrix} \left\langle |k_1|^2 \right\rangle & 0 & 0 \\ 0 & \left\langle |k_2|^2 \right\rangle & 0 \\ 0 & 0 & \left\langle |k_3|^2 \right\rangle \end{bmatrix}. \tag{2.3}$$

As explained in Chap. 4, the $[A]$ matrix (which keeps the same information of $\underline{k}$) performs the weighting of the two vectors $\underline{\omega}_T$ and $\underline{\omega}_P$ separately.

When the formulation follows a physical approach, the $[P]$ matrix is derived by the covariance matrix (Boerner 2004; Cloude 1987; Zebker and Van Zyl 1991) expressed in the basis which makes $\underline{\omega}_T = [1,0,0]^T$:

$$[C] = \left\langle \underline{k}\, \underline{k}^{*T} \right\rangle = \begin{bmatrix} \left\langle |k_1|^2 \right\rangle & \left\langle k_1 k_2^* \right\rangle & \left\langle k_1 k_3^* \right\rangle \\ \left\langle k_2 k_1^* \right\rangle & \left\langle |k_2|^2 \right\rangle & \left\langle k_2 k_3^* \right\rangle \\ \left\langle k_3 k_1^* \right\rangle & \left\langle k_3 k_2^* \right\rangle & \left\langle |k_3|^2 \right\rangle \end{bmatrix}. \tag{2.4}$$

A. Marino, *A New Target Detector Based on Geometrical Perturbation Filters for Polarimetric Synthetic Aperture Radar (POL-SAR)*, Springer Theses, DOI: 10.1007/978-3-642-27163-2, © Springer-Verlag Berlin Heidelberg 2012

The mathematical justification of deleting the off-diagonal terms is associated with the bias removal for the polarimetric coherence estimation. In fact, correlation between target and clutter can bias the detector leading to false alarms and missed detections.

In both algebraic and physical approaches the detector exploits the diagonal matrix $[P]$ where the off-diagonal (or cross) terms are neglected. The aim of this appendix is to prove that the operation is proper and useful information is not lost (in the context of single target detection). The main concern against neglecting the off-diagonal terms could be that without the second order statistics (i.e. cross terms) a partial target cannot be completely characterised (Cloude 1992). Therefore, a partial target could not be separated by a single target constituting false alarms.

The demonstration will be divided in two parts. Firstly, we want to prove that the cross terms of $[C]$ are not needed to detect a single target and secondly that the algorithm developed can deal with partial target clutter. Instead than providing a single proof we preferred to collect several proofs exploiting different aspects of the problem, in order to present different points of view.

2.2 Uniqueness for Single Target Detection

The doubt regarding the uniqueness raises since the detector appears to be constructed with three power terms. The latter are merely three real numbers, while a single target has five degrees of freedom (five real numbers). As it will be proven shortly, this is deceptive since the required parameters are hidden inside the final formulation.

To summarise we want to demonstrate:

the algorithm is able to detect uniquely any single target.

The thesis could be articulated more in details:

excluding the off-diagonal (cross) terms, the three diagonal (power) terms are sufficient to grant uniqueness to the detection of single targets unless a small dispersion in the geometrical target space.

2.2.1 Number of Degrees of Freedom Exploited

In the new basis the scattering mechanisms for target and perturbed target are respectively $\underline{\omega}_T = [1, 0, 0]^T$ and $\underline{\omega}_P = [a, b, c]^T$, while the scattering vector is $\underline{k} = [k_1, k_2, k_3]^T$. The final detector as derived in Chap. 4 is

$$\gamma_d = \frac{1}{\sqrt{1 + \frac{|b|^2}{|a|^2} \frac{P_{C2}}{P_T} + \frac{|c|^2}{|a|^2} \frac{P_{C3}}{P_T}}}, \qquad (2.5)$$

where the power terms are calculated as $P_T = \left\langle |k_1|^2 \right\rangle$, $P_{C2} = \left\langle |k_2|^2 \right\rangle$ and $P_{C3} = \left\langle |k_3|^2 \right\rangle$.

An easy way to estimate the power terms is by the Gram-Schmidt ortho-normalization (Strang 1988; Hamilton 1989; Rose 2002), which sets $\underline{\omega}_T$ as one axis of the new basis of the target space. The new basis will be composed of three unitary vectors $\underline{u}_1 = \underline{\omega}_T$, $\underline{u}_2 = \underline{\omega}_{C2}$ and $\underline{u}_3 = \underline{\omega}_{C3}$, where again $\underline{\omega}_{C2}$ and $\underline{\omega}_{C3}$ are two components orthogonal to $\underline{\omega}_T$ (lying on the clutter complex plane which is orthogonal to the target complex line). Hence, P_T, P_{C2} and P_{C3} can be calculated with the squared amplitude of the inner products (i.e. projections) between the observables $\underline{k}$ and the three axes of the basis

$$P_T = \left\langle \left| \underline{k}^T \underline{\omega}_T \right|^2 \right\rangle, \quad P_{C2} = \left\langle \left| \underline{k}^T \underline{\omega}_{C2} \right|^2 \right\rangle \text{ and } P_{C3} = \left\langle \left| \underline{k}^T \underline{\omega}_{C3} \right|^2 \right\rangle \qquad (2.6)$$

Any scattering mechanism can be represented by 4 parameters (e.g. Huynen or α model) (Cloude and Pottier 1997; Huynen 1970). If the dependence on the Huynen parameters is explicated the powers will be written as

$$P_T(\phi_m, \tau_m, \gamma, \upsilon) = \left\langle \left| \underline{k}^T \underline{\omega}_T(\phi_m, \tau_m, \gamma, \upsilon) \right|^2 \right\rangle,$$
$$P_{C2}(\phi_m, \tau_m, \gamma, \upsilon) = \left\langle \left| \underline{k}^T \underline{\omega}_{C2}(\phi_m, \tau_m, \gamma, \upsilon) \right|^2 \right\rangle, \qquad (2.7)$$
$$P_{C3}(\phi_m, \tau_m, \gamma, \upsilon) = \left\langle \left| \underline{k}^T \underline{\omega}_{C3}(\phi_m, \tau_m, \gamma, \upsilon) \right|^2 \right\rangle.$$

The detector is based on the power terms estimated starting from the scattering vector and mechanisms which were demonstrated to be necessary and sufficient to characterise any single target. The power terms are the result of the projections of the scattering vector and not the starting point. In other words, we use all the information contained in the scattering vector to estimate the power terms (i.e. five parameters instead than three).

2.2.2 Rank of the Covariance Matrix

Any single target can be interpreted as lying on a subspace of the entire partial target space (Cloude 1986, 1995b). Specifically, a partial target with pure polarisation is by definition regarded as single. However, single and partial target are frequently treated separately, as they would be two completely different entities.

In order to show the narrow link between single and partial targets, the incoherent eigenvalue decomposition (Cloude 1987, 1992; Cloude and Pottier 1996) could be considered. When a single target is present in the averaging cell, only one eigenvalue differs from zero. The diagonal matrix has rank one:

$$[\Sigma] = \begin{bmatrix} \lambda_1 & 0 & 0 \\ 0 & 0 & 0 \\ 0 & 0 & 0 \end{bmatrix}. \tag{2.8}$$

Clearly, also the covariance matrix $[C]$ has rank one, since it can be represented as the product of the same (or parallel) scattering vectors over the entire cell, hence its columns are dependent on each other.

In general, any single target constitutes a subspace with covariance matrix of rank one. Consequently, in order to characterise uniquely a single target, a rank one covariance matrix is necessary and sufficient. In $[\Sigma]$, the cross terms are clearly zero (as well as the other two diagonal terms), therefore in the basis which diagonalise the covariance matrix they are not needed to characterise uniquely the single target. The diagonal expression $[\Sigma]$ for the covariance matrix is obtained only after a change of basis where the single target represents one axis (the first one). This change of basis is achieved with a similarity transformation, i.e. left and right multiplication for a unitary matrix composed of the eigenvectors (Cloude 1992).

In the proposed algorithm, the change of basis is imposed in the first step, subsequently the obtained covariance matrix is made diagonal. The algorithm can be interpreted as a test for the fit of the imposed diagonal matrix with the data. Clearly, if in the data there is only the sought single target, the changes of basis for diagonalisation and detector will generate the same diagonal matrix. In the latter case the match will be high and the target will be detected.

In conclusion, a diagonal matrix (specifically of rank one) is necessary and sufficient to characterise a single target after an appropriate change of basis (Cloude 1992).

2.2.3 Test of Uniqueness and Target Dispersion

The span of the scattering matrix can be calculated with the *Trace* of the covariance matrix (i.e. sum of the diagonal terms). It represents the total power acquired by the receiving antenna in a quad polarimetric mode (Mott 2007; Ulaby and Elachi 1990). It represents a physical property of the target, therefore it is invariant on changes of basis. In other words, $Trace\{[C]\}$ will remain the same independently on the basis used to express $[C]$. Therefore, the basis which makes $\underline{\omega}_T = [1, 0, 0]^T$ can be taken into account.

Additionally, it is apparent that

$$Trace\{[C]\} = Trace\{[P]\}, \tag{2.9}$$

since off diagonal terms do not contribute in the *Trace*.

After the change of basis the span can be expressed as

$$SP = P_T + P_{C2} + P_{C3}, \tag{2.10}$$

where SP is the span.

In order to prove the uniqueness, we will consider two different targets $\underline{k}_{T1}$ (target of interest) and $\underline{k}_{T2}$ (test target) and demonstrate that the targets can be detected simultaneously if and only if they are the same, unless a small dispersion. Starting from the expression of the coherence

$$\gamma = \frac{1}{\sqrt{1 + RedR\left(\frac{SP}{P_T} - 1\right)}} \geq T \Rightarrow \frac{1}{T^2} \geq 1 + RedR\left(\frac{SP}{P_T} - 1\right)$$

$$\frac{1}{RedR}\left(\frac{1}{T^2} - 1\right) \geq \frac{SP}{P_T} - 1 \tag{2.11}$$

$$1 \leq \frac{SP}{P_T} \leq 1 + \frac{1}{RedR}\left(\frac{1}{T^2} - 1\right)$$

In the last expression, the left inequality is true because the span is always (equal or) higher than the power scattered by a target.

$$1 \leq \frac{P_T + P_C}{P_T} \leq 1 + \frac{1}{RedR}\left(\frac{1}{T^2} - 1\right), \tag{2.12}$$

$$0 \leq \frac{P_C}{P_T} = \frac{1}{SCR} \leq \frac{1}{RedR}\left(\frac{1}{T^2} - 1\right). \tag{2.13}$$

The final expression is dependent on the ratio between target and clutter: SCR.

After the change of basis, a single target of interest can be expressed as $\underline{k}_{T1} = [\sigma e^{j\varphi_1}, 0, 0]^T$. This is different from the scattering mechanism which is $\underline{\omega}_T = [1, 0, 0]^T$, moreover the phase φ_1 can be arbitrary since it cannot be used to characterise the target (Cloude 2009). It is always possible to express a scattering vector as

$$\underline{k}_{T2} = \left[(\sigma + \Delta\sigma')e^{j\left(\varphi_1' + \Delta\varphi_1'\right)}, \Delta\sigma_2'e^{j\varphi_2'}, \Delta\sigma_3'e^{j\varphi_3'}\right]^T. \tag{2.14}$$

$\underline{k}_{T2}$ represents the test target in our proof.
The detectors for target and test targets are:

$$\underline{k}_{T1} : 0 = \frac{1}{SCR_1} = \frac{0}{\langle\sigma^2\rangle} \leq \frac{1}{RedR}\left(\frac{1}{T^2} - 1\right), \tag{2.15}$$

$$\underline{k}_{T2} : 0 \leq \frac{1}{SCR_2} = \frac{\left\langle(\Delta\sigma_2')^2\right\rangle + \left\langle(\Delta\sigma_3')^2\right\rangle}{\left\langle(\sigma + \Delta\sigma')^2\right\rangle} \leq \frac{1}{RedR}\left(\frac{1}{T^2} - 1\right). \tag{2.16}$$

The target $\underline{k}_{T1}$ will be always detected since its signal to clutter ratio is ∞.
As shown by the last set of equations, the only way for the two targets to have exactly the same detector is to have the same SCR:

$$0 = SCR_1 = SCR_2,$$

$$\frac{0}{\langle \sigma^2 \rangle} = \frac{\left\langle \left(\Delta\sigma_2^I \right)^2 \right\rangle + \left\langle \left(\Delta\sigma_3^I \right)^2 \right\rangle}{\left\langle \left(\sigma + \Delta\sigma^I \right)^2 \right\rangle} \Rightarrow \left\langle \left(\Delta\sigma_2^I \right)^2 \right\rangle + \left\langle \left(\Delta\sigma_3^I \right)^2 \right\rangle = 0. \quad (2.17)$$

In other words, the clutter components of the test target must be zero and the test target can be expressed in the polarimetric space with a vector parallel to the target (i.e. they are the same single target).

However in Eq. 2.16 an inequality is considered, therefore a dispersion of the test target which still allows its detection is quantified. Any target falling in the dispersion equation will be detected. In order to have a more direct picture of the polarimetric information of the detectable targets, a normalisation of the scattering vector is performed. In actual fact, the detector is not dependent on the norm of the scattering vector, hence we do not lose generality restricting our analysis to normalised targets.

A normalised vector has unitary length:

$$\sqrt{\left(\sigma + \Delta\sigma^I \right)^2 + \left(\Delta\sigma_2^I \right)^2 + \left(\Delta\sigma_3^I \right)^2} = 1. \quad (2.18)$$

For the test target we have:

$$\left\langle \left(\sigma + \Delta\sigma^I \right)^2 \right\rangle = 1 - \left\langle \left(\Delta\sigma_2^I \right)^2 \right\rangle + \left\langle \left(\Delta\sigma_3^I \right)^2 \right\rangle, \quad (2.19)$$

$$\frac{\left\langle \left(\Delta\sigma_2^I \right)^2 \right\rangle + \left\langle \left(\Delta\sigma_3^I \right)^2 \right\rangle}{1 - \left\langle \left(\Delta\sigma_2^I \right)^2 \right\rangle + \left\langle \left(\Delta\sigma_3^I \right)^2 \right\rangle} \leq \frac{1}{RedR} \left(\frac{1}{T^2} - 1 \right) = x. \quad (2.20)$$

where x is a positive real number.

$$\left\langle \left(\Delta\sigma_2^I \right)^2 \right\rangle + \left\langle \left(\Delta\sigma_3^I \right)^2 \right\rangle \leq \frac{x}{1 + x}. \quad (2.21)$$

Now, we want to find a simplified expression for $\dfrac{x}{1 + x}$:

$$\frac{x}{1 + x} = \frac{1}{RedR} \left(\frac{1}{T^2} - 1 \right) \Big/ \left(1 + \frac{1}{RedR} \left(\frac{1}{T^2} - 1 \right) \right) = \quad (2.22)$$

$$\frac{1 - T^2}{RedR \cdot T^2} \Big/ \frac{RedR \cdot T^2 + 1 - T^2}{RedR \cdot T^2} = \frac{1 - T^2}{1 + T^2(RedR - 1)}. \quad (2.23)$$

Substituting the found expression the dispersion equation becomes:

$$\left\langle \left(\Delta\sigma_2^I \right)^2 \right\rangle + \left\langle \left(\Delta\sigma_3^I \right)^2 \right\rangle \leq \frac{1 - T^2}{1 + T^2(RedR - 1)}, \quad (2.24)$$

where clearly, T and $RedR$ are positive.

The final expression defines the maximum spreading of the clutter components of a target to be still detected. In other words, any normalised target with clutter components smaller than the dispersion boundaries is detected.

The dispersion equation is determined by the threshold and the *RedR*. In order to have a deeper understanding of the relation between dispersion and parameters, the limits can be calculated (Riley et al. 2006):

$$\lim_{tr \to 1} \frac{1 - T^2}{1 + T^2 (RedR - 1)} = 0. \tag{2.25}$$

Hence an extremely restrictive threshold (i.e. $T = 1$) will reduce the dispersion to zero.

$$\lim_{tr \to 0} \frac{1 - T^2}{1 + T^2 (RedR - 1)} = 1. \tag{2.26}$$

An extremely low threshold (i.e. $T = 0$) allows the detection of any target even when the target component is inexistent. Please note, 1 is the maximum value for the clutter components since the vector is normalised.

$$\lim_{RedR \to 0} \frac{1 - T^2}{1 + T^2 (RedR - 1)} = 1. \tag{2.27}$$

When the *RedR* is zero and the clutter terms are infinitely reduced, the dispersion becomes maximum and any target can be detected. Considering, $RedR = \frac{|b|^2}{|a|^2}$ and the scattering mechanism is unitary, the only way to have $RedR = 0$ is with $|b| = 0$ and $|a| = 1$, hence $\omega_T = \omega_P$. In other words, we are considering the normalised inner product of a vector for itself, which is always unitary.

The last limit is:

$$\lim_{RedR \to \infty} \frac{1 - T^2}{1 + T^2 (RedR - 1)} = 0. \tag{2.28}$$

In order to have $RedR = \infty$ the perturbed target must have only clutter components (it must be on a complex plane orthogonal to the target of interest). Hence,

$$\omega_T \perp \omega_P = [0, b, c]^T \tag{2.29}$$

and the normalised inner product of two orthogonal vectors is always zero (Strang 1988).

In conclusion, the higher is the value for the *RedR*, the more restrictive is the filter.

Regarding the best choice for the parameters, in order to perform detection on real data, the dispersion must be small but not zero, since the observed targets do not fulfil perfectly the models (at least for the thermal noise introduced by the instrument). Additionally, the presence of surrounding clutter must be included in the dispersion, and a *SCR* of interest must be chosen.

To have an idea about the amount of dispersion allowed in a practical detection, we can substitute the values of *RedR* and *tr* which were used in the validation chapter. If $RedR = 0.25$ and $T = 0.97$ we have $\left(\Delta\sigma_2''\right)^2 + \left(\Delta\sigma_2''\right)^2 < 0.20$. This means that the first component is around 0.8. In term of angles distances in the power space, the dispersion allows detection of targets 14 degrees far from the target axis. We regard this variation sufficiently small, however in the case a more selective filter is required the value of *RedR* and *T* can be adjusted as appropriate.

2.3 Detection in Scenario Populated by Partial Targets

With the collection of proofs provided in the previous section, we demonstrated that the detection of a single target is unique and we can neglect the off diagonal terms of [C] (after the change of basis). Additionally, a dispersion equation was extracted starting from the threshold and the Reduction Ratio.

In this section, we want to prove:

when the detection is applied on a real scenario (i.e. in presence of partial targets), the off diagonal (cross) terms of [C] can still be neglected without decreasing the detection performances.

Geometrically, the proof is rather straightforward. Conversely than a single target, a partial target cannot be described with a rank one covariance matrix (Cloude 1992). The operation of neglecting the cross terms (without changing the diagonal terms) is not a similarity and it generally can modify the rank of the matrix. However, if the initial matrix is a covariance matrix, the diagonal matrix [P] will not reduce its rank Please note, the rank can increase but in presence of the target the matrix will already be diagonal (with only the first element different from zero) and it will not be affected. In order to prove this last property we can consider by absurd a covariance matrix in any basis (for instance the basis which makes $\underline{\omega}_T = [1, 0, 0]^T$) expressed as:

$$[\tilde{C}] = \langle \underline{k}\,\underline{k}^{*T} \rangle = \begin{bmatrix} \langle |k_1|^2 \rangle & \langle k_1 k_2^* \rangle & \langle k_1 k_3^* \rangle \\ \langle k_2 k_1^* \rangle & 0 & \langle k_2 k_3^* \rangle \\ \langle k_3 k_1^* \rangle & \langle k_3 k_2^* \rangle & 0 \end{bmatrix}. \tag{2.30}$$

The $[\tilde{C}]$ matrix has generally rank 3 (since its determinant is generally different from zero) however the reduced diagonal matrix [P] has rank one. Therefore, if such matrix would exist, to neglect the off diagonal terms would lead to false alarm since some partial target would be interpret as single. Evidently, $[\tilde{C}]$ does not represent a physically realisable target (Cloude 1986; Mathews and Howell 2006, Riley et al. 2006). The only way to have diagonal terms equal to zero:

$P_{C2} = \left\langle |k_2|^2 \right\rangle = 0$ and $P_{C3} = \left\langle |k_3|^2 \right\rangle = 0$ is for $k_2 = k_3 = 0$ over the entire averaging cell. However, this would lead to $\langle k_1 k_2^* \rangle = \langle k_1 k_3^* \rangle = \langle k_2 k_3^* \rangle = 0$. The latter relationship can be seen as a consequence of the Cauchy-Schwarz inequality (Strang 1988) where

$$\langle |k_i| \rangle \langle |k_j| \rangle \geq \left| \left\langle k_i \, k_j^* \right\rangle \right|. \tag{2.31}$$

Moreover, it is not positive semi-definite (Rose 2002; Strang 1988). The only way for a covariance matrix to have two zeros on the diagonal (i.e. all the minors constructed using those columns will be zero) is to be rank one. In this situation, there exists a basis where the projection of the partial target power over two axes is zero. The two axes will span a complex plane where the partial target must be always zero. Therefore, in this basis the target power will be present only on one axis, which is the definition of single target. Generalising, partial targets cannot be always zero over any complex plane in the single target space $SU(3)$ (Cloude 1986). This can be related to the presence of a pedestal representing the unpolarised component which spreads all over the target space.

In conclusion, neglecting the cross terms, we do not lose information regarding the partial nature of the target, since the partial target will always have the other two diagonal elements. Clearly, the detector is not able to discriminate between two partial targets since it cannot characterise completely partial targets. However, the algorithm is able to understand when a target is partial and discard it.

References

Boerner WM (2004) Basics of radar polarimetry. RTO SET Lecture Series

Cloude RS (1992) Uniqueness of target decomposition theorems in radar polarimetry. Direct Inverse Methods Radar Polarim 1:267–296

Cloude SR (1987) Polarimetry: the characterisation of polarisation effects in EM scattering. Electronics engineering department. University of York, York

Cloude RS (1995a) An introduction to wave propagation antennas. UCL Press, London

Cloude SR (1995b) Lie groups in EM wave propagation and scattering. In: Baum C, Kritikos HN (eds) Electromagnetic symmetry. Taylor and Francis, Washington, pp 91–142. ISBN 1-56032-321-3

Cloude SR (2009) Polarisation: applications in remote sensing. Oxford University Press, Oxford, 978-0-19-956973-1

Cloude SR, Pottier E (1996) A review of target decomposition theorems in radar polarimetry. IEEE Trans Geosci Remote Sens 34:498–518

Cloude SR, Pottier E (1997) An entropy based classification scheme for land applications of polarimetric SAR. IEEE Trans Geosci Remote Sens 35:68–78

Cloude SR (1986) Group theory and polarization algebra. OPTIK 75:26–36

Hamilton AG (1989) Linear algebra: an introduction with concurrent examples. Cambridge University Press, Cambridge

Huynen JR (1970) Phenomenological theory of radar targets. Delft Technical University, The Netherlands

Lee JS, Pottier E (2009) Polarimetric radar imaging: from basics to applications. CRC Press, Boca Raton

Mathews JH, Howell RW (2006) Complex analysis for mathematics and engineering. Jones and Bartlett, London

Mott H (2007) Remote sensing with polarimetric radar. Wiley, Hoboken

Pearson CE (1986) Numerical methods in engineering and science. Van Nostrand Reinhold Company, New York

Riley KF, Hobson MP, Bence SJ (2006) Mathematical methods for physics and engineering. Cambridge University Press, Cambridge

Rose HE (2002) Linear algebra: a pure mathematical approach. Birkhauser, Berlin

Strang G (1988) Linear algebra and its applications, 3rd edn. Thomson Learning, New York

Ulaby FT, Elachi C (1990) Radar polarimetry for geo-science applications. Artech House, Norwood

Zebker HA, Van Zyl JJ (1991) Imaging radar polarimetry: a review. Proc IEEE 79:1583–1606

Printed by Publishers' Graphics LLC
CAMZ140302.20.06.208